The Palgrave Lacan Series

Jacques Lacan is one of the most important and influential thinkers of the 20th century. The reach of this influence continues to grow as we settle into the 21st century, the resonance of Lacan's thought arguably only beginning now to be properly felt, both in terms of its application to clinical matters and in its application to a range of human activities and interests. The Palgrave Lacan Series is a book series for the best new writing in the Lacanian field, giving voice to the leading writers of a new generation of Lacanian thought. The series will comprise original monographs and thematic, multi-authored collections. The books in the series will explore aspects of Lacan's theory from new perspectives and with original insights. There will be books focused on particular areas of or issues in clinical work. There will be books focused on applying Lacanian theory to areas and issues beyond the clinic, to matters of society, politics, the arts and culture. Each book, whatever its particular concern, will work to expand our understanding of Lacan's theory and its value in the 21st century.

More information about this series at
http://www.palgrave.com/gp/series/15116

Isabel Millar

The Psychoanalysis of Artificial Intelligence

palgrave
macmillan

Isabel Millar
Centre for Critical Thought
University of Kent
Canterbury, UK

The Palgrave Lacan Series
ISBN 978-3-030-67980-4 ISBN 978-3-030-67981-1 (eBook)
https://doi.org/10.1007/978-3-030-67981-1

Cover illustration: VICTOR HABBICK VISIONS/SCIENCE PHOTO LIBRARY/gettyimages

This Palgrave Macmillan imprint is published by the registered company Springer Nature Switzerland AG.
The registered company address is: Gewerbestrasse 11, 6330 Cham, Switzerland

This book is dedicated to my mum, Sylvia.

Prologue: Roko's Basilisk

In 2010 on *LessWrong* forum, a user named Roko posited a thought experiment. He proposed that in a hypothetical future an all-powerful super-intelligent AI could retrospectively punish anyone who in the present time did not do everything in their power to aid in the creation of such a superintelligence. By merely entertaining the idea of such a being and not facilitating its development you would expose yourself to the possibility that it would deduce that you had not acted in accordance with the duty to bring it into existence (the moralistic tone of the experiment is enforced by the fact that the AI is paradoxically a benevolent one whose task is to protect humankind, and therefore those who don't facilitate its existence desire ill against their fellow men). The vengeful Abrahamic nature of the Basilisk meant that in future, it could recreate a simulation of you to torture for all eternity for the sin of putting him at existential risk. The Old Testament stylings of the Basilisk are clear: *he*'s nice, but only if you deserve it.

As absurd as the tale sounds, it was met with outrage by the site's founder and director of the *Machine Intelligence Research Institute* (MIRI) in California, Eliezer Yudkowsky. Yudkowsky felt that Roko had opened a pandora's box of previously unimaginable torment that the poor readers of his blog would now fall victim to. In response to Roko's post he reportedly said:

> Listen to me very closely, you idiot.
>
> YOU DO NOT THINK IN SUFFICIENT DETAIL ABOUT SUPERINTELLIGENCES CONSIDERING WHETHER OR NOT TO BLACKMAIL YOU. THAT IS THE ONLY POSSIBLE THING WHICH GIVES THEM A MOTIVE TO FOLLOW THROUGH ON THE BLACKMAIL.
>
> You have to be really clever to come up with a genuinely dangerous thought. I am disheartened that people can be clever enough to do that and not clever enough to do the obvious thing and KEEP THEIR IDIOT MOUTHS SHUT about it, because it is much more important to sound intelligent when talking to your friends.
>
> This post was STUPID (ibid.).[1]

The post was subsequently removed, and all talk of the Basilisk was banned from the website for over five years. But the Basilisk had already wreaked havoc among the forum's readers many of whom had started to experience psychological difficulties. Paranoiac fears of the Basilisk's future existence have now become something between an urban legend and a genuine topic of philosophical debate, not to mention the fact that it is taken seriously by some of the major tech entrepreneurs and scientists currently driving AI research. The logic behind the Basilisk is even (spuriously) backed up by Timeless Decision Theory and Bayesian probability.

In fact, Yudkowsky (2010) has written at length on the theory underpinning the problem of the Basilisk, even drawing on the prisoner's dilemma which we will recall Lacan (2006a) uses in his discussion of logical time. The prisoner's dilemma was a thought experiment in game theory, where the actions of several prisoners were dependant on the anticipated decisions of one another in order for them to secure their freedom. The dilemma exemplified for Lacan the tripartite structure of time surreptitiously at work in the concept of so-called rational thought. These he called the instant of seeing, the time for understanding, and the

[1] See David Auerbach (2014).

moment of concluding. Accordingly, whilst logical time is not objective, this does not mean that it cannot be formulated according to a rigorous structure; that of intersubjective logic based on a dialectical relation between hesitation and urgency. A logic we see at work in Roko's autopoietic Basilisk and what could be called in other terms, *hyperstition*.

The term hyperstition was coined by Warwick University's *Cybernetic Cultural Research Unit* (CCRU) and continues to be one of the major concepts of the Accelerationist movement. A portmanteau of 'hyper' and 'superstition', drawing on the Baudrillardian logic of hyperreality, hyperstition to paraphrase Nick Srnicek and Alex Williams (2014), the authors of the *#Accelerate Manifesto for Accelerationist Politics*, refers to narratives capable of bringing themselves into reality through the workings of feedback loops, which generate new socio-political attractors. Roko's Basilisk allegedly functions according to just this sort of hyperstitous logic. As a computational form of Pascal's wager, it relies on a number of premises for it to function. Firstly, the proviso that the concept of a Singularitarian superintelligence entails the capacity for absolute and total recall of all data and secondly the ability to simulate every historically living being in order to then torture them. Thirdly, the belief that a simulation is equivalent to a subject. As Ana Teixera Pinto (2018) has noted however, the theological and paranoiac overtones of the Basilisk function as:

> the personification of AI as Oedipal beast [...] and of code as the male seed. Those who seek mathematical proof of the prediction's likelihood are missing the point. The content of Roko's thought experiment is symbolic, not scientific: it speaks through cipher and allegory. (p. 19)

Teixera Pinto highlights the Oedipal logic at work in the positing of the Basilisk, but to this we might add that the phallic enjoyment involved in the imagining of the ultimate mathematizable *One* that admits of no exemptions is masculine logic par excellence. The Basilisk also functions as the ultimate indicator of anxiety, the impossible object as cause of desire and also complete destruction. The poor human on this score is trapped between the finite slab of meat that tortures him and the infinite simulation that he will inevitably become. What seems to be at stake in this speculation on the Singularity is what Lacan (2006b) in *Function and Field* referred to as the future anterior:

> What is realized in my history is neither the past definite as what was, since it is no more, nor even the perfect as what has been in what I am, but the future anterior as what I will have been, given what I in the process of becoming. (p. 247)

In the logic of Roko's Basilisk, we may apprehend the möbius structure of the relationship between AI and psychoanalysis that this book will attempt to depict, a topology which for as far as one travels along, will always lead inevitably to its inverse: its *extimate* core.

References

Auerbach, D (2014) The Most Terrifying Thought Experiment of All Time. Available (01.03.2020) at: https://slate.com/technology/2014/07/rokos-basilisk-the-most-terrifying-thought-experimentof-all-time.html

Lacan, J. (2006a) 'Logical Time and the Assertion of Anticipated Certainty' in J. Lacan, *Écrits*, pp. 161–175. London: W.W. Norton & Company.

Lacan, J. (2006b) 'The Function and Field of Speech and Language in Psychoanalysis' in *Écrits*, pp. 197–268. London: W.W. Norton & Company.

Srnicek, N., & Williams, A. (2014) '#Accelerate Manifesto for an Accelerationist Politics' in N. Srnicek & A. Williams (eds.) *#Accelerate: The Accelerationist Reader*, pp. 347–362. Falmouth: Urbanomic.

Teixera Pinto, A. (2018) The Psychology of Paranoid Irony. *Transmediale Journal* 1: pp. 18–22.

Yudkowsky, E. (2010) *Timeless Decision Theory. The Machine Intelligence Research Institute*. Available (01.03.20) at: https://intelligence.org/files/TDT.pdf

Praise for *The Psychoanalysis of Artificial Intelligence*

"Does It think? Does It enjoy? Taking the problem of artificial intelligence as a problem that has been in a way always-already inherent to psychoanalytic inquiry Isabel Millar accomplishes a most powerful and productive shift of perspective on both psychoanalysis and AI. Her work takes us on a fascinating journey across a vivid conceptual and figural landscape, and provides an excellent proof that powerful, captivating theory is all about asking the right kind of questions. *The Psychoanalysis of Artificial Intelligence* is both extremely timely and timeless in the way it constructs and tackles its object."

—Professor Alenka Zupančič, *The European Graduate School and Slovenian Academy of Sciences and Arts*. Author of *Ethics of The Real: Kant and Lacan and What is Sex?*

"Boldly drawing on a vast range of academic disciplines orchestrated by an enviable psychoanalytic erudition and an original treatment of so-called "sexbots" as a central object for contemporary speculative and social investigation, Millar's book asks a series of seminal and long-overdue questions, which are here to stay. How should we approach the allegedly forthcoming advent of the "singularity" in terms of sexuality and sexuation? Does sexual reproduction have a future? What new forms of enjoyment, if any, might Artificial Intelligence enable us to think and experience? Or is it rather the case that androids secretly already have wet dreams about the human-all-too-human absence of the sexual relationship?"

—Dr Lorenzo Chiesa, *Newcastle University, UK*. Author of *Subjectivity and Otherness* and *The Not-Two*

"People tend to respond to artificial intelligence with either fear or love. Isabel Millar proposes a third way: to psychoanalyze artificial intelligence and the persistent investment in it. In a stunning work of expansive intellectual power, Millar shifts the fundamental question concerning artificial intelligence to the terrain of enjoyment. After Millar's book, the question "Does it enjoy?" should be the starting point for any engagement with artificial intelligence. It is simply an epochal book for understanding this engagement."

—Professor Todd McGowan, *Department of English, University of Vermont, USA*. Author of *The Real Gaze and Emancipaton After Hegel*

Contents

List of Figures

1
Introduction

The percentage of intelligence that is not human is increasing. And eventually, we will represent a very small percentage of intelligence.
—Elon Musk (2018, online)

The Psychoanalysis of Artificial Intelligence, what a strange proposition. What could it possibly mean? The significance of the two terms in themselves is hardly self-evident, let alone their relationship to one another. Psychoanalysis on the one hand; simultaneously a clinical practice, a mode of cultural critique and a philosophical battle ground. And Artificial Intelligence, a technoscientific 'invention' originating in the 1950s[1] yet with literary, cultural and fantasmatic origins that date back centuries, and a concept whose theoretical potential continues to provoke intense philosophical debate. In this book, I argue that Artificial Intelligence (AI) and the creation of the artificial brain, which promises to separate neuroscience from biology and thought from the body, along with the prospect

[1] The earliest coinage of the term Artificial Intelligence is attributed to computer and cognitive scientist John McCarthy at a 1956 workshop at Dartmouth College Other attendees at the workshop, who would soon become founders and leaders in the early field of AI research, were Allen Newell CMU, Herbert Simon (CMU), Marvin Minsky (MIT) and Arthur Samuel (IBM).

I. Millar, *The Psychoanalysis of Artificial Intelligence*, The Palgrave Lacan Series,
https://doi.org/10.1007/978-3-030-67981-1_1

of forms of embodied AI which aim to simulate and surpass human intelligence, provokes an urgent engagement with the psychoanalytic subject. Simultaneously the book considers psychoanalysis as a crucial tool in our understanding of what AI means for us as speaking, sexed subjects. In short, AI and psychoanalysis stand in *extimate* relation to one another.

Through the reconceptualization of Intelligence, the Artificial Object and the Sexual Abyss we conjure a figure who exists on the boundary of psychoanalysis and AI, straddling our fantasy worlds and our speculations about the possibilities for life alongside or through Artificial Intelligence; the Sexbot. With its help, and through the medium of film we subvert Kant's three famous enlightenment questions, *What Can I Know, What Should I Do and What May I Hope For*. Ultimately, we transition from the question *can it think* to *does it enjoy?*

Owing to its inherent conceptual interdisciplinarity it is no wonder that AI and the discourses surrounding it seem to have unique capacity to blur the boundary between science and fiction. Embedded in a rich history of fantasy and pop-science, elements of which have been the subject of philosophical reflection since antiquity, appearing in various guises throughout the history of Western thought and literature,[2] it is often difficult to discern where the science of AI starts and fiction ends. Today there is no unifying theory which guides Artificial Intelligence research given that it draws from a variety of fields including, computer science, information theory, mathematics, neurobiology, psychology, linguistics, logic and analytic philosophy. Its potential and scope are in constant debate both scientifically and conceptually, being a polemical topic for cultural theory, political thought, ethics, philosophy, and even cosmology. Considering the rapid advances made regarding the reverse-engineering of the human brain in the field of neural networks and deep learning and the adjacent fields of quantum computing, nano and biotechnology, some, like futurist Ray Kurzweil (2014) anticipate that we will soon transcend the "limits of nature", thereby reaching a synthesis of science and fiction in the 'Singularity'. Others argue we are about to enter

[2] We may recall for example Ovid's *Pygmalion,* Descartes (alleged) robotic daughter Francine (Kang 2017), Maelzel's chess playing automaton and Čapek's (2004) *Rossun's Universal Robots*, to name but a few instances.

a "Fourth Industrial Revolution": an era heralding the gradual fusion of digital, physical, and biological worlds (Schwab 2016). For many philosophers and theorists of AI this so-called *Life 3.0* (Tegmark 2017) where science-fiction becomes terrifying reality is a conceptual terrain, which raises complex questions about the notion of intelligent life, the nature of thinking, the future of the social bond and the constitution of the "human". In *Superintelligence, Paths, Dangers and Strategies*, Nick Bostrom (2014), foresees a dark future for humanity if we ignore his warnings about the possibility of a Hal 9000-like artificial Superintelligence, by which he means any intelligence that vastly exceeds the performance of humans. He believes that the creation of a super intelligent being could lead to the extinction of humankind. The risk involved in the creation of Superintelligence is that it would be operating on a speed and scale unfathomable to humans, which could initiate an intelligence explosion on a digital time scale of millisecond speed so powerful as to accidentally (or deliberately) destroy humanity. Bostrom not only contemplates the possibility of malicious applications of AI, such as hacked military devices, nano-factories distributed in undetectable concentrations creating killing devices on command and even payed human 'dupes' doing AI's dirty work but envisions a scenario in which, once AI achieves a stage of world domination, humans would be useful only as raw materials. As he puts it: 'brains, if they contain information relevant to the AI's goals, could be disassembled and scanned, and the extracted data transferred to some more efficient and secure storage format' (Bostrom 2014, p. 118). In order to prevent the emergence of such rogue Superintelligences, Bostrom joined Stephen Hawking in 2015 to sign an open letter on behalf of *The Future of Life Institute*, warning of the possible threats of AI. The signatories all subscribed to twentythree principles to ensure the safe development of Artificial Intelligence. As Max Tegmark (2017) enumerates, however, there are many misconceptions and disagreements about the future of AI. These include questions of when, how and what form AI will take and how long the process of its evolution will be. Furthermore, the possibility of so-called Superintelligence is still highly contested. This, however, has not prevented some from speculating about the possible date of its arrival. The Singularity designates just this hypothetical moment of an intelligence explosion, a point-of-no-return, at which AI will decisively surpass human intelligence, rendering the human species as we know it obsolete, if not actually extinct.[3] Its foremost advocate, Kurzweil, expects the Singularity will occur in two phases. By 2029 AI will supposedly reach the stage of human level or 'General' Artificial Intelligence and successfully pass a Turing Test and in 2045 humankind will multiply its effective intelligence by a billion-fold through merging with AI. The potentially paradigm shifting consequences of the hypothetical emergence of general or super-intelligent AI has even become a topic for cosmology. Veteran scientist and inventor of the Gaia Hypothesis James Lovelock recently published *Novacene* (2019) in which he proposes that the age of the Anthropocene (the geological period in which humans acquired planetary scale technology) has already come to an end and we are entering a new age, the 'Novacene' in which technology will come to inherit the 'consciousness' of the cosmos. In his vision, artificially intelligent beings who can think 10,000 times faster than humans will emerge as the inheritors of the earth and caretakers of the intelligent universe. For Lovelock, the hypothesis of the emergence of such intelligent beings makes it even more vital that we retain the environmental conditions conducive to their survival. Thus, as Yuval Harari (2017) observes, the central hallmark of debates on the future of AI is the hubristic question: 'Who are the new "Gods"—humans or AI?'

This concern with the technological Singularity as some sort of onto-theological watershed moment is taken up by Žižek (2020) who remarks that, what the advocates of the Singularity often fail to realise, or at least fully engage with, is that in this passage from human to post-human, what disappears is precisely self-awareness, which is rooted in 'finitude and failure' (p. 75). Regarding the apparent paradox which emerges as a result of our popular visions of post-human Singularity, Žižek goes on to state that:

[3] The term was popularised by science fiction writer Victor Vinge in 1983 and brought into wider circulation by his (1993) article '*The Coming Technological Singularity*'. According to David Chalmers (2010), however, the term Singularity is used in a variety of ways to refer to different scenarios; the loose sense refers generally to the unpredictable consequences of exponential growth in AI, while the Singularity in the strict sense refers to a point where 'speed and intelligence go to infinity' (p. 3).

> Insofar as posthumanity is, from our finite/mortal human standpoint, in some sense the point of the Absolute towards which we strive, the zero-point at which the gap between thinking and acting disappears, the point at which I became *homo deus*, we encounter here again the paradox of our brush with the Absolute: the Absolute persists as the virtual point of perfection in our finitude, as that X we always fail to reach, but when we get over the limitation of our finitude we also lose the Absolute itself. Something new will emerge, but it will not be creative spirituality relieved of mortality and sexuality—in this passage to the new we will definitely lose both. (p. 158)

Whilst Žižek's diagnosis of the problem with discourses on the Singularity is apposite, here my concern will not be to repeat the same gesture but rather to seek a constructive and productive way to engage with our relationship to AI psychoanalytically. While I will not attempt to give an account of the historical development of (or philosophy of) Artificial Intelligence, I will delineate a general working definition of Artificial Intelligence as: *a non-human mode of thought, whether embodied or disembodied, which acts autonomously and whose motives and purpose we may not necessarily be aware of, nor even understand.* Some might say that conveniently this definition could also be applied to the psychoanalytic conception of the unconscious, an ambivalence that lies at the heart of this book. Recall in *Seminar II* Lacan's (1988) reproach to Octave Mannoni for his worries over the human becoming too much like a machine:

> Don't be soft. Don't go and say that the machine is really rather nasty and that it clutters up our lives. That is not what is at stake. The machine is simply the succession of little 0s and 1s, so that the question as to whether it is human or not is obviously entirely settled—it isn't. Except, there's also the question of knowing whether the human, in the sense in which you understand it, is as human as all that. (p. 319)

Between 1985 and 1986, at the psychoanalysis department at the *Université Paris 8*, Jacques-Alain Miller gave his course on *Extimité* in which he characterized the logic of the Lacanian unconscious as an extroverted interiority. 'Extimacy', a portmanteau of exterior and intimate, is a word first coined by Lacan (1992) in *The Ethics of Psychoanalysis.*

Although Lacan did not explicitly return to the concept in any of his seminars, the logic of extimacy, following Miller (1988), can be said to underpin the Lacanian organon in general as a concern with the intimate exteriorization that belies the nature of subjectivity, most clearly articulated in Lacan's relentless concern with the topological coordinates of the möbius strip, the Klein bottle and knot theory. However, not only is the unconscious qua 'discourse of the Other' (1988, p. 89) to be understood in terms of an extroverted interiorization that morphs the notion of "unconscious depth" into a question of topological space, but as this book attempts to illustrate, the very materiality of the *speaking body* in its relation to Artificial Intelligence should be understood as extimate.

In a civilization in which Artificial Intelligence is becoming a significant element in the social bond, the psychoanalysis of AI is a provocation. It asks us to question both the meaning of psychoanalysis when taken outside of the purview of the strictly 'human' clinical space and conversely it attempts to show in what ways psychoanalysis is already an extimate part of artificial intelligence. Similarly, it speculates on what form our philosophical and critical thinking about AI has hitherto neglected the essential element or indeed material of psychoanalysis, that is to say, enjoyment. This leads us to proffer the hypothesis that the 'vanishing mediator' between our two unlikely bedfellows is none other than sex. For psychoanalysis and its clinical treatment of 'suffering', sex is the crucial problem underlying all others. But more than a symptomatic 'problem' sex is a philosophical problem. Philosophical in the sense that it has, by definition, no solution. For psychoanalysis sex names the impossible yet inevitable collision of epistemological and ontological questions that characterize the entrance into subjectivity for all speaking beings. So, we must ask, what is sex for Artificial Intelligence? Judging by most of the literature and popular discourse surrounding it, sex is nothing more than an apparently superficial anthropomorphization of our fantasies of AI. But isn't this precisely the point? This fantasy of AI sex obscures the fact that sex is *only ever* a fantasy covering up for a hole in reality itself, or in Baudrillardian terms a question of dissimulation as a strategy of simulation. It is an absence which, as this book hopes to illustrate, brings with it a deafening silence which is impossible to ignore. The 'sex' of Artificial Intelligence resides everywhere, it is what brings it into being. In Lacanian terms we could qualify this further to say that AI in its many forms both actual and fantasmatic ex-ists as a form of relation to the signifier or more specifically, a mode of enjoyment. Through the employment of both the philosophical engagements and the clinical and conceptual developments of Lacanian theory, the book aims to develop a novel and productive encounter between psychoanalysis and AI. In proposing to approach Artificial Intelligence psychoanalytically the book views the sexual non-rapport as its theoretical kernel. I seek firstly to advance a psychoanalytic reading and problematization of AI as a discourse about 'knowledge in the real'. Secondly, to develop a novel, conceptual grid to query the material implications of Artificial Intelligence for subjectivity, the body and the social bond. In this sense, this project is not concerned with simply providing a psychoanalytic elucidation of our unconscious fears, fantasies or fascination with AI. Rather, it seeks to take the real dimensions of AI seriously. In short, this means the passage from a concern with the barred-subject and *object a* to a concern with the speaking body and the *artificial object*; one which Lacan in *Seminar XVII* gave the provisional name *lathouse*. The lathouse is an under-theorized and underutilised Lacanian concept, which presents us with a new way of understanding our bodily and structural relationship to AI.

So how *does* one read the sentence which forms the title of this book? Are we planning to psychoanalyze AI? If so, what would that mean? Or are we inquiring after the possibility of AI to be the psychoanalyst? This begs questions of how we are to conceptualize of AI as a 'thinking thing'. The first ambiguity we should draw attention to however is the fact that psychoanalysis strictly speaking only ever happens as the result of a demand, a subjective and singular demand on the part of the analysand. And this demand is met with the desire of the analyst, for whom the demand of the analysand, is an *object a*. Both these essential elements give rise to a transference relation resulting in what could be characterized as psychoanalysis proper. The wager of this book is that paradoxically in order to understand the stakes of Artificial Intelligence it is not to post-humanism or transhumanism that we should turn but rather to the subversive spirit and (anti-humanism) of Lacanian psychoanalysis, taking the 'demand' of AI as our *object a*.

In 1973 Jacques-Alain Miller interviewed Jacques Lacan for a French television broadcast in which he challenged the renegade psychoanalyst about the nature and value of his psychoanalytic theory and practice. Lacan's responses were typically elliptical, but nonetheless provide the careful reader with an encrypted summary of his work to date and the place of Lacanian psychoanalysis in the contemporary world. Interestingly, Miller's interview concludes with his positing of the three Kantian questions to Lacan: '*What can I know*?', '*What ought I to do*?' and '*What may I hope for*?' Lacan offers Miller short shrift in response owing to what, in his view, is the difference in the role of the psychoanalyst as opposed to the philosopher. Perhaps the key to his reply can be found several pages earlier where he refers to the function of the Saint as corresponding to the '*trashitas*' of society (1990, p. 15); a position which, he says, must be taken up by the psychoanalyst as the '*refuse of jouissance*' (p. 16). It is not, in Lacan's view, for the analyst to ask the Kantian questions, but rather to allow the subject to realise his position with respect to them. The fourth Kantian question '*What is man?*' was never broached in this interview, but one could argue that it constitutes the underlying thread that runs through the whole of the psychoanalytic edifice.

I will therefore revisit the three Kantian questions which Miller challenged Lacan to address in the 1970s in the new context of Artificial Intelligence and via the prism of sexual non-rapport. The Kantian questions, which defined the Enlightenment project, will be employed to examine and problematize the relationship between psychoanalysis and AI. The three questions are typically present in all popular discourse and critical speculations on the future of AI. The first usually with reference to the question of consciousness and the perennial problem of "other minds". This is articulated in concerns with the sentience of Artificial Intelligence, perhaps most famously exemplified by the Turing Test as the ultimate "measurement of consciousness". The second Kantian question characteristically revolves around the ethics of AI; to what extent do we allow various forms of AI to enter into the social bond and how do we prevent its worst excesses or impacts on us as subjects? The third Kantian question is centred on the notion of the Singularity. Will we need to contemplate a future living with other forms of intelligence? Or will the advent of Superintelligence signal the end of humanity and thus the extinction of the species as we know it? While the book poses the Kantian questions à la Miller, it refuses to answer them, à la Lacan. Instead of the standard approach taken by most philosophers or critical theorists on the problems of AI, I will look rather for the other side (*l'envers*) of the questions.

So, what forms of AI will this book be concerned with? AI is as huge and complex an object of scrutiny as psychoanalysis, and this project can by no means cover the entirety of either of those domains. My more modest task is to clarify a manner in which the two realms would find each other's extimate kernel residing inside themselves. In order to do this, I have conjured a conceptual figure who exists on the boundary of psychoanalysis and AI. To this end, the first part of the book is concerned with providing the theoretical groundwork for the conceptualisation of the Sexbot via a psychoanalytic examination of the concept of intelligence, the artificial object and the abyssal nature of sex.

Once I have drawn up this figure, I turn to the speculative work of the book in the form of the three Kantian questions. I mobilise the Sexbot as a *figure* to articulate the ontological, epistemological and technological series of problems that underly the entrance of AI into the social bond. The figure of the Sexbot, as represented in its ideal form in film, is to be understood as the *sinthome*[4] which binds together AI, the sexual non-rapport and the *lathouse*. The Sexbot as a theoretical device attempts to address the impossibility of the sexual relation for speaking beings, in the sense of the necessity of a supplement to cover up the void of sex and at the same time the inevitability of the problem of sex for Artificial Intelligence. Through the metonymy of the Sexbot as exterior, interior and finally extimate in relation to the subject or speaking body the book will address the various dimensions of the psychoanalysis of AI. Given the speculative nature of this project I have chosen somewhat counterintuitively to use the medium of film to address these dimensions. However, it should be clarified that whilst I engage with film, I do not read film

[4] In line with Lacan's later work, (specifically *Seminar XXIII*) the symptom is replaced by the sinthome; the precise configuration of elements (imaginary, symbolic and real) which constitute the regime of enjoyment for any speaking body. Used in this context, the concept of the sinthome represents the tripartite unification of disparate dimensions inextricably held together by a common thread.

itself as a medium.[5] In other words, here the films function as a conceptual playground to explore the modes of enjoyment inherent to the psychoanalysis of AI within the theoretical framework of the Sexbot. Kant's questions will be contextualised according to the new conceptual concerns relating to Artificial Intelligence and its problematization of the sexual non-relation. The films discussed therefore are chosen for their ability to illustrate the different aspects of the psychoanalysis of AI as epitomised by the signifiers Knowledge, Act and Hope. This will inevitably lead us to engage with Kant's fourth question: '*What is man?*'

Ultimately the crucial concept running through the book is enjoyment or jouissance. Jouissance here is thought of not merely as a supplement to subjectivity but its essential component, it is what structures thought itself. On this score masculinity and femininity pertain not just to gender identities but to forms of abstract thought which may be employed as a framework for analysing (or indeed psychoanalysing) Artificial Intelligence. It is therefore the concept of jouissance and its fundamental relationship to knowledge that articulates the transition from the traditional philosophical concern about AI as 'can it think?', to the psychoanalytic concern, 'does it enjoy?' And if so, the question we are left with remains: is there something new about this AI enjoyment that goes beyond our previous models of masculine and feminine subjectivity as abstract modes of thought? Lacan (1998), whilst not talking about artificial intelligence perhaps sums this up with the following enigmatic statement:

> Man believes he creates—he believes believes, believes, he creates creates, creates. He creates creates, creates woman. In reality, he puts her to work—to the work of the One [...]. That is what S(A̸) means. It is in that respect

[5] Whilst psychoanalytic film theory in its traditional incarnations will not be employed, it must be acknowledged that through the work primarily of Todd McGowan, the field of Lacanian film theory has taken a turn closer to matching the goals of this project. In the sense that the more recent invocations of Lacan for film analysis engage less with the question of the spectator, the audience and the cinematic experience *per se* and more with the structural and conceptual mechanisms of film as a mode of speculative thought For McGowan (2007) where traditional film theory had located the gaze on the side of the spectator, this in his view was a fundamental misreading of Lacan. The gaze for McGowan following Lacan's meaning of the term should be located outside of the subject as an intrusive presence which emanates from an unseen place, accordingly the gaze is the invisible space within the filmic image itself.

that we arrive at the point of raising the question how to make the One into something that holds up, that is, that is counted without being. Mathematization alone reaches a real [...] a real that has nothing to do with what traditional knowledge has served as a basis for, which is not what the latter believes it to be—namely, reality—but rather fantasy. The real, I will say, is the mystery of the speaking body, the mystery of the unconscious. (p. 131)

Bibliography

Bostrom, N. (2014) *Superintelligence: Paths, Dangers, Strategies*. Oxford: Oxford University Press.

Čapek, K. (2004) *R.U.R. (Rossum's Universal Robots)*. London: Penguin Books.

Chalmers, D. (2010) The Singularity: A Philosophical Analysis. *Journal of Consciousness Studies* 17(9–10): pp. 7–65.

Harari, Y.N. (2017) *Homo Deus: A Brief History of Tomorrow*. New York: HarperCollins Publishers.

Kang, M. (2017) The Mechanical Daughter of Rene Descartes: The Origin and History of an Intellectual Fable. *Modern Intellectual History* 14(3): pp. 633–660.

Lacan, J. (1988) *The Seminar of Jacques Lacan Book II: The Ego in Freud's Theory and in the Technique of Psychoanalysis 1954–1955*. London: W.W. Norton & Company.

Lacan, J. (1990) *Television: A Challenge to the Psychoanalytic Establishment*: London: W.W. Norton & Company.

Lacan, J. (1992) *The Seminar of Jacques Lacan Book VII: The Ethics of Psychoanalysis*. London: W.W. Norton & Company.

Lacan, J. (1998) *The Seminar of Jacques Lacan Book XX: Encore—On Feminine Sexuality, the Limits of Love and Knowledge 1972–1973*. London: W.W. Norton & Company.

Lovelock, J. (2019) *Novacene: The Coming Age of Hyperintelligene*. London: Penguin.

McGowan, T. (2007) *The Real Gaze: Film Theory After Lacan*. New York: SUNY Press.

Miller, J-A. (1988) Extimité. *Prose Studies* 11(3): 121–31

Schwab, K. (2016) *The Fourth Industrial Revolution*. London: Penguin Random House.
Tegmark, M. (2017) *Life 3.0: Being Human in the Age of Artificial Intelligence*. London: Penguin.
Vinge, V. (1993) The Coming Technological Singularity: How to Survive in the Post-Human Era. *Lewis Research Center, Vision 21: Interdisciplinary Science and Engineering in the Era of Cyberspace*: pp. 11–22.
Žižek, S. (2020) *Sex and the Failed Absolute*: London: Bloomsbury.

Part I

2

The Stupidity of Intelligence

As soon as intelligence takes itself as its object, it is destined to transform into stupidity, either as the g factor or as intellect. If the psychologists' intelligence is stupid, then, in the end, that of the philosophers may be equally so. The philosophical self-assertion of the mind, claiming the sovereignty of the mind or intellect, always seems to result in a ridiculous form of celebration of the self that is no better than the reductionism of psychologists.
—Catherine Malabou *(2019, pp. 51–52)*

2.1 AI: Failure, Trauma, Dupery

The dictionary definition of intelligence, from the Latin 'intus' (between) and 'legere' (to choose), entails variously the ability to discern, make decisions, comprehend, to have skill, art, taste and ultimately knowledge.[1] But the term 'artificial' preceding 'intelligence' gives one the immediate

[1] Indeed, the question of the genealogy of the concept of intelligence has a rich history in the fields of philosophy, psychology and cognitive science, the literature on which I shall not delve into here. Suffice it to say, once we interrogate the concept of intelligence, very quickly, we are in the realms of philosophy and the humanities in general and not science (much less computer science) *stricto sensu*.

I. Millar, *The Psychoanalysis of Artificial Intelligence*, The Palgrave Lacan Series,
https://doi.org/10.1007/978-3-030-67981-1_2

impression that we have a sound definition of what real intelligence is; a conceit that in philosophical terms elides *technē* and *epistēmē* and in psychoanalytic terms takes for granted the distinct categories of knowledge, truth, and most significantly, enjoyment.

Human intelligence entails various interactions between different skills, for example, a combination and interaction of visual perception, motor skills, memory, speech, spatial reasoning, auditory processing may be utilised at any given moment. These skills are of course not all transparently understandable to the 'intelligent human' utilizing them. This is the paradox at the heart of debates between neuroscience and philosophical accounts of consciousness[2] which start from fundamentally different premises on how we may talk about subjective phenomena. For example, at its crudest level, just because you can see, it does not mean that you 'know' how vision works. And conversely, knowing how vision works does not guarantee that you will be able to see. This same sort of combination of functions will be present in any complex AI program, which will have integrated elements the core processor knows how to access. This may include evidence-based reasoning, language skills, text analysis, sensors, decision making, data analysis and so on.

For example, can a program that processes visual information be said to understand how vision works? Or does facial recognition software operate in the same intuitive way for computers as it does for humans? According to some computer scientists the answer is yes, computers like humans have "intuition" (Hammond 2018). So, in very basic terms we can see that there is a discrepancy between technical or machinic capacity and theoretical knowledge, which is integral to the concept of Artificial Intelligence as it is commonly used. However, as Lydia Liu (2010) notes, it is one thing to argue that the human brain can be augmented by the computer chip and another to imagine that the logic of the computer and

[2] Classically exemplified by John Searle's (1980) *Chinese Room Argument*; a thought experiment in which a hypothetical language processing machine would be constructed by a non-Chinese speaking person inside a room manipulating Chinese characters according to a set of rules and outputting them to another Chinese-speaking person outside the room. Searle used the example to critique Strong AI on the basis that the person inside could not be said to understand Chinese despite being able to convince its interlocuter otherwise. The argument has been refuted since in various ways, objecting that metaphorically speaking the person inside the room may logically correspond to a part of the brain but not consciousness *per se*.

the human psyche are analogous, this prosthetic argument, as she calls it, is an 'alibi for something more fundamental that has been going on since the mid-twentieth century, and this is the cybernetic conception of the human psyche as a computing machine' (p. 8). What remains to date the pervasive paradigm for thinking about Artificial Intelligence is the relationship between the computer and the brain.[3] But as we shall see this specific focus on the analogy of the brain and the computer leaves important psychoanalytic questions unbroached.

In recent times, the field of Artificial Intelligence has grown exponentially with a multitude of technical approaches provoking interest from various philosophical perspectives surrounding its conceptual ramifications. Today, AI encapsulates a huge range of phenomena including those that fall under goal-oriented "narrow" AI or Artificial Narrow Intelligence (ANI), which perform limited tasks such as sorting, tracking, predicting or recognising data patterns based on a range of different algorithms. These may be used in applications like Google Search and Amazon's Alexa, or more sophisticated AI in the development of autonomous vehicles or AVs (Apple's Project Titan and Tesla's Autopilot), preventative medicine (Microsoft AI) and, more controversially, autonomous weapon systems. At the other end of the spectrum, however, one finds robotics, deep learning and neural networks, which find a more complex outlet in simulations and perhaps most intriguingly, the Blue Brain Project as we shall discuss.

A recent volume of essays *Alleys of Your Mind: Augmented Intelligence and its Traumas* tackles the question of Artificial Intelligence from the perspective of contemporary critical theory.[4] The collection gathers contributions from leading thinkers who all in diverse ways criticise popular conceptions of AI and ask the questions; what thinking means in the age of Artificial Intelligence, and how does big-scale computation transform the ways in which our brains function. The book's main conceit is to

---

[3] The equation, or rather the evolutionary development of the human brain and the computer in neuroscientific research and computer science first emerged as a result of John Von Neumann's (2012) contributions to the fields of information technology and computation, best captured in his 1958 *The Computer & The Brain*. See also Kurzweil (2012).

[4] The authors gathered in this collection are predominately philosophers or new media theorists but include also historians of science and art theorists.

uncover the 'positive role played by error and trauma in the construction of our contemporary technological minds' (Pasquinelli 2015).

In his contribution to the volume, Matteo Pasquinelli (2015) argues that current philosophical debates on the question of technology and reason fall between, on the one hand, neomaterialism and, on the other, neorationalism (or in other words, positions that draw on either Whitehead or Sellars). The former being a proponent of the 'agency of technical objects, matter and affects' and the latter addressing rather the 'primacy of reason and its potential forms of autonomization' (p. 8), or the ability for synthetic reason to become autonomous. Pasquinelli sees this as the wrong distinction to be making however, and believes instead that no paradigm of cognition and computation can be assessed without the recognition of the 'epistemic abnormal' or what he calls 'noetic failure' which we could translate as mental or intellectual error (p. 8). This would mean a distinction should be made between philosophies that acknowledge a positive role for 'error, abnormality, pathology, trauma and catastrophe' and those who support a flat ontology without these 'dynamic, self-organizing and constitutive ruptures on the other' (ibid.). Following the Frankfurt School's lessons on the trauma of reason, Pasquinelli asserts that the *reason of trauma* must be 'rediscovered as the actual inner logic of the age of intelligent machines' (ibid.). In his introduction to the volume, he asserts that:

> One day, it will not be arbitrary to reframe twentieth century thought and its intelligent machines as a quest for the positive definition of error, abnormality, trauma and catastrophe—a set of concepts that need to be understood in their cognitive, technological and political composition. (p. 7)

Drawing parallels with Foucault's history of biopower and technologies of the self as sharing common roots with cybernetics and its 'error-friendly machines' (p. 7) and arguing that Deleuze and Guattari's desiring machines were in fact echoing research on war trauma and brain plasticity from the First World War, he states that:

> Across the history of computation (from early cybernetics to Artificial Intelligence and current algorithmic capitalism) both mainstream technol-

> ogy and critical responses to it have shared a common belief in the determinism and positivism of the *instrumental* or *technological rationality* to use the formulation of The Frankfurt School. (Ibid.)

The aim of the collection therefore is to highlight conversely the role of 'error, trauma and catastrophe in the design of intelligent machines and the theory of augmented cognition' (p. 7). Pasquinelli argues that the definition of intelligence remains an open problem since from a philosophical point of view, human intelligence is always artificial in the first place, engendering as it does novel dimensions of cognition (ibid.). Intelligence is a combination of various complex and multifaceted capacities, which defy easy definition. Pasquinelli discerns three major "fallacies" in the current debates on the state of Artificial Intelligence. Firstly, the anthropocentric fallacy, which naively assumes Artificial Intelligence would be analogous to human intelligence in the superficial sense of attributing menacing and threatening motives to AI. In this vision of AI, it is a malevolent predator whose aim is to expunge us or at the very least make us suffer. Secondly, the bootstrapping fallacy, which imagines a seamless exponential growth of machine intelligence similar to the progress of human psychological development, in the sense that the relative and progressive complexity of cognitive tasks for humans is directly mapped onto the progress of machinic intelligence. This fallacy errs when it fails to reckon with the different forms of cognitive capability that human intelligence entails, and the ways in which some processes that are very difficult for humans are easily replicable by algorithms while other seemingly simple human tasks are hugely complex engineering feats for AI. And finally, the third problem is the Singularity fallacy, which, combining elements of the previous two problems, entails the belief that there will be a decisive point of unification and synchronisation of different technologies simulating, augmenting and ultimately surpassing human intelligence which taken together form a homogenous and all-powerful mode of thought, capable of rendering human intelligence (and, by extension, the species in general) redundant.

Pasquinelli's criticisms of the current fallacies surrounding conceptions of AI and of the need to approach AI from the basis of error, trauma, and catastrophe, points to the implicitly psychoanalytic dimensions and implications of AI. Firstly, the issue Pasquinelli points to with imaginary identification, which seems to be pervasive in our misapprehensions of Artificial Intelligence. This entails the assumption that the ability to communicate symbolically entails an element of intersubjectivity, which is fundamentally misguided. In Lacanian terms this relates to the question of misrecognition or imaginary *méconnaisance* and, ultimately, transference. Secondly, we should consider more closely the characterization of the concept of trauma and pathology as the unacknowledged ontologically significant factor in the positivist project of AI research. This is the concern that the whole of the Freudian enterprise was built on; the concept of trauma and the re-formulation of what was previously conceived as pathology as inherent to psychic structure. This was not least because it was via trauma and the symptom that Freud found the way into deciphering his very first patient's hysterical symptoms, and neither merely in the sense of the discovery of an unfortunate and disturbing event, but rather that trauma, as Freud described it (and Lacan formalised it), was structurally constitutive of the psychoanalytic subject. The subject, in psychoanalytic terms, is characterised by a constitutive failure.

Benjamin Bratton (2015) likewise warns that contemporary polemics around the meaning of thinking and living with radically other forms of synthetic intelligence crucially misunderstand the real issues at stake, stuck as they are in anthropocentric appreciations of AI. As far as Bratton is concerned, we should resist the temptation to understand AI through the lens of human intelligence, for to do so would be 'self-defeating, unethical and perhaps even dangerous' (p. 70). For this reason, he advocates a broadening up of the concept of intelligence to situate human intelligence as just one form of intelligence along a larger continuum. In this way he promotes a form of "anti-bigotry" in our understanding of what may count as intelligent thought; a step towards a better appreciation of the challenges involved in living and thinking with the Other of Artificial Intelligence.

Bratton suggests that our fantasies of AI as either desperately wishing to be human e.g. Steven Spielberg's *A.I.* (2001) or Chris Columbus' *Bicentennial* (1999) *Man* or malevolently fixated on our destruction e.g. James Cameron's (1984) *The Terminator* (1984) or more recently, various episodes of Charlie Brooker's series *Black Mirror* merely reflect our own desires, paranoia and narcissistic self-image. That AI would be so invested in getting our attention is perhaps pure wishful thinking; the worst-case scenario may well be that they wouldn't even notice us at all:

> the real nightmare, even worse that the one in which the Big machine wants to kill you, is the one in which it sees you as irrelevant, or not even as a discrete thing to know. Worse than being seen as an enemy, is not being seen at all. (p. 70)

The idea of Artificial Intelligence being perceived as like humans, Bratton. says, is a valid point of departure, but not a valid conclusion to end up with. Furthermore an important distinction Bratton makes in AI research is between the terms artificial stupidity and artificial idiocy, where the first term refers to faults deliberately programmed into AI in order to make it more realistically human (i.e. not winning every time in a game in which it could very easily beat a human every time). The second term refers to the problem that occurs when an AI performs its tasks too well, to the detriment of other factors. The apocalyptic example used originally by Nick Bostrom (2014), is the paperclip generator that follows its orders to continue making paperclips until the world is overrun by an avalanche of paper clips. This is the idiocy of AI so called. In his first example, AI stupidity has the function of creating a social bond between itself and its human companion. AI idiocy on the other hand has the aim of following its master's instructions so "ideologically" that it allows all other factors to pail into insignificance. So here the blind spot functions in different places. Or one could say the enjoyment produced is different in each case. The stupidity or idiocy of AI as Bratton calls it may therefore be seen as a discursive structure, in which enjoyment is produced according the relative positions of truth and knowledge. The concept of stupidity therefore is one which has fundamental structural importance in relation to the question of enjoyment.

Bratton also draws attention to what he sees as the inherently 'bigoted' approach humans have to AI, since as he (2015) puts it, the Turing Test's aim was to fool an interlocutor that an AI was human. So the AI had to be 'in drag' (p. 76). But as Bratton remarks, the thing about actual drag is, you are not supposed to convince someone you are a member of the opposite sex, but merely get them to suspend their belief about your sexual or gender identity (in the disavowed form of 'I know very well, but…'). This he discusses with relation to the scandal surrounding Turing's outing as a homosexual man (illegal at the time) and his subsequent chemical castration, causing Turing an unspeakable suffering which in the end led to him commit suicide. Given that Turing's Imitation Game was based on the logic of tricking an interlocutor into believing that you were a particular gender (at which the AI had to perform as well as the human) the parallels are significant.

> One notes the sour ironic correspondence between asking an AI to pass the test in order to qualify as intelligent—to pass as human intelligence—with Turing's own need to hide his sexuality and to pass as a straight man. The demands of both bluffs are unnecessary and profoundly unfair. (p. 72)

However, the question of the relationship between sexual identity and AI is in fact far more psychoanalytically complex than a political irony about the potential problems with anthropocentrism in AI research as Bratton suggests. As we shall discus further in Chap. 5, the Turing Test, when closely examined, demonstrates the fundamental basis for the relationship between sexuation in psychoanalytic terms and Artificial Intelligence.

In essence, Bratton's arguably utopian message is that instead of forcing AI into prefigured stereotypes about the type of thinking that we may recognize as "real" we should allow AI in all its diverse forms, about which we are continually learning, to teach humans a more 'fuller and truer range of what thinking can be' (ibid., p. 72). Diversifying what counts as "thought" is in keeping with the realist turn in contemporary theory and offers us an important perspective in relation to pushing psychoanalytic paradigms beyond post-structuralist tendencies. But furthermore we should pay attention to the fact that (as Bratton himself points out in his example of AI stupidity) there is a constitutive part of "thought" that involves a dialectical relationship between the positions of truth and knowledge in any given discursive framework.

2.2 Omega Numbers and Suture

Let us press further the question of thinking in philosophical approaches to AI. In her essay '*Instrumental Reason, Algorithmic Capitalism and the Incomputable*' Luciana Parisi (2015) asks:

> Can the critique of instrumental rationality—as addressed by critical theory—still be based on the distinction between critical thinking and automation? Can one truly argue that algorithmic automation is always already a static reduction of critical thinking? (p. 126)

She argues that with the all-machine phase transition of digital capitalism, we are bearing witness to a new mode of thought and control. This phase identified by a group of physicists from the University of Miami, coincides with the introduction of high frequency stock trading after 2006 and entails sub-millisecond speed algorithm to algorithm interactions that exceed human response time. Having analysed the millisecond scale data at the core of financial markets, they discovered a series of sub second extreme events caused by these algorithms. Given this state of affairs, Parisi argues that due to the inhuman scale on which these events unfold, this changes significantly our capacity to analyse them using the tools of traditional critical theory. Crucially she asserts that the traditional critical theory reproach that accuses computation of reducing human thought to mechanical operations is no longer sufficient as an analytical paradigm for our current state of affairs. Since these events are outside the limits of human control and comprehension, she cites computer scientist and mathematician Gregory Chaitlin's conviction that incomputability and randomness are in fact the very condition of computation. This means that the incomputable forms part of instrumental rationality itself. Parisi identifies the dilemma thus:

> Both philosophical thought and digitality, rely on principals of indetermination and uncertainty while featuring these principles in their core complexity theories. As such, both challenge and define the neoliberal order at the same time—a paradox. (p. 126)

Addressing this paradox, she argues for a turn to Chatlin's concept of incomputability, and specifically his discovery of the incomputable 'Omega number'; a number which is definable, but not computable. As Parisi explains: 'Omega defines at once a discrete and an infinite state of computation occupying the space between zero and one' (p. 126). She sees this as bringing not only the philosophical critique of technical rationalization into question but also the instrumentalization of reason. This signals, she says, an 'irreversible transformation in the history of critical thought in which the incomputable function of reason has entered into the automated infrastructure of cognition' (p. 127).

However, this recognition from a Lacanian point of view is actually not so new at all. Parisi's identification of the mysterious Omega number which unsettles all attempts to ground rationality on computation, being both a discrete yet infinite space between zero and one, and therefore incomputable, uncannily brings to mind the fundamental psychoanalytic notion of suture. In Jacques-Alain Miller's (2012) '*Suture*' paper, 'the first great Lacanian text not to be written by Lacan himself' (Badiou 2008, p. 25) Miller presents a logic of the signifier, now widely accepted as the strict formalization of the *Lacanian* logic of the signifier, although never systematized by Lacan himself. The paper grounds the logic with recourse to Frege's attempts at a logical conception of the sequence of whole natural numbers in which the categories of concept, object and number are distinguished in order to establish the zero as the only object subsumable under the concept of 'non-identity to itself'. As the editors of *Concept and Form* note, two relations are furthermore assumed by Miller: 'the subsumption of the object under the concept' and the 'assignation of a number to the concept' (Cahiers Kingston 2012, online).

By virtue of the thing non identical with itself being evoked only to then be excluded from the dimension of truth, Miller (2012) shows how Frege grounds the concept of number on a performative contradiction:

> This concept, by virtue of being a concept, has an extension, subsumes an object. *Which object? None.* Since truth is, no object falls into the place of the subsumed of this concept, and the number which qualifies its extension is zero. In this engendering of the zero, I have stressed that it is supported by the proposition that truth is. If no object falls under the concept of

> non-identical-with-itself, it is because truth must be saved. If there are no things which are not identical with themselves, it is because non-identity with itself is contradictory to the very dimension of truth. To its concept, we assign the zero. (p. 97)

Via this reading of Frege, Miller elaborates the paradoxical logic of the signifier and the subject of lack present in all discourses that aim at truth. Without entering into a comparative set-theoretical or conceptual discussion of Parisi's position versus Frege, it suffices to highlight that significant critical approaches to the question of AI seem to coincide with a logical paradox about the position of truth within discourse. In Parisi's case the supposed discrepancy that lies between the opposed discourses of rationality and computation are undermined by the apparition of a mysterious *stain on reality* that does not seem to fit into either side of the binary opposition of subjectivity and objectivity as she characterizes it. In reference to the Turing Machine qua 'absolute mechanism of iteration based on step by step procedures' (p. 130), Parisi argues that:

> Nothing is more opposed to pure thought—'or the being of the sensible' as Deleuze (1994: 68) called it—than this discrete—based machine of universal calculation. The Turing architecture of pre-arranged units that could be interchangeably exchanged along a sequence is effectively the opposite of an ontogenetic thought moving through a differential continuum, through intensive encounters and affect. (pp. 130–131)

What does Parisi here mean by the Deleuzian reference to ontogenetic thought, intensive encounters and affect? How does this translate into psychoanalytic terms?

This concern with the mysterious Omega number we may also read with reference to Lacan's early engagements with cybernetics in *Seminar II* and his (2006a) *écrit* '*Logical Time and the Assertion of Anticipated Certainty*' in which he draws attention to the significance of the algorithmic calculation of possibilities central to cybernetic theory and its possible use for constructing a psychoanalytic theory of causation. In his essay on logical time, Lacan illustrates via the thought experiment of the prisoner's dilemma, how rational calculation also contained within it a temporal dimension that was entirely subjective and conversely how free action is only ever constituted via a universal rule:

> What makes this act so remarkable in the subjective assertion demonstrated by the sophism is that it anticipates its own certainty owing to the temporal tension with which it is subjectively charged; and that, based on this very anticipation, its certainty is verified in a logical precipitation that is determined by the discharge of this tension—so that in the end the conclusion is no longer grounded on anything but completely objectified temporal instances, and the assertion is desubjectified to the utmost. (Lacan 2006a, p. 171)

When deciding how to act in order to secure their freedom, the prisoners were entirely dependent on their intersubjective positioning towards one another, this involved not the flat ontology of hyper-rationality but, as Wright (2018) observes, the temporality of 'living scansion' (p. 75). The remarkable thing here is that the any act of freedom retroactively constitutes a completely determined mode of causality but is nonetheless experienced as subjectively authored. *In Seminar II* Lacan uses Edgar Allen Poe's *Purloined Letter* to illustrate the fundamentally illusory effect of chance that resides in the game of odds and evens, highlighting how, we tend to see coincidence in certain numerical patterns where mathematically speaking there is only probability; chance being purely an effect of structure. This becomes significant clinically when in *Seminar XI* Lacan makes an important distinction between two different kinds of psychoanalytic causality broadly corresponding to necessity and contingency. The first Lacan calls *automaton* and refers to a type of programmed repetition of behavior, and the second, *tuché*, corresponds to the chance and unpredictable encounter with the real. In terms of the speech of the analysand this is experienced as an unprecedented event, a radical subjective freedom that erupts from of the chain of pure machinic determination by the symbolic order. It is this unpredictable encounter with the real that we may call 'stupid' in that it by necessity lacks any rational explanation. It seems that Parisi envisions rationality as precisely this continuum of immanent affect, totally transparent to itself and self-positing, in other words *being* and *thinking* here are the same thing. Parisi diagnoses the problem as one of discrete units versus continuous movement, perhaps a problem one could put differently in Lacanian terms as tuché versus automaton (Lacan 1977), or mathematization versus topology (Lacan 2016)?

Whilst Parisi attempts to obviate the question of the subject in relation to AI, it nevertheless slips in via the back door. For what is the function of the Omega number if not to guarantee absolute contingency in necessity and absolute freedom within determinism? It seems the Omega number, like the Lacanian real, *always returns to its place* and *never stops not being written*.

2.3 Malabou and the Blue Brain

In 2005 the Blue Brain Project was set up by Professor Henry Makram at the EPFL (2020), the Swiss Federal institute of technology in Lausanne, Switzerland. The aim of the project is to 'build biologically detailed digital reconstructions and simulations of the rodent brain and ultimately the human brain'. The project's super-computer based reconstructions promise to offer a radically new approach to understanding and simulating the multilevel structure and functioning of the brain. Whilst the progress of the project continues a pace, due to the immense complexity of the human brain, the BBP still only manages to simulate tiny portions of the brain in order to extrapolate larger findings from their limited but ever-growing harvest of data. The aim, according to their mission statement, however, is to be able to simulate in ever more detail the complex multi-level activity across different parts of the brain that would be impossible to investigate in living tissue. Researchers at the BBP may manipulate tissue in various ways, for example by 'knocking out' or 'lesioning' parts of the circuit.

As it stands, the researchers are capable of digitally reconstructing brain tissue 'in silico' to simulate a snapshot of the anatomy and physiology of the brain at any one moment in time. They can use these digital reconstructions for a virtually unlimited range of simulations in order to replicate the spontaneous electrical activity of a real brain. The project is currently building neurorobotic tools, which will enable the researchers to replicate cognitive behavioral experiments in animals, with the aim of extrapolating this data to generate further understanding and simulations of the human brain. Ultimately, BBP is attempting what has hitherto been the preserve of science fiction and fantasy; building a fully functional non-biological brain. Even though the technology needed is still in its early stages, in theory with enough computer power, the Blue Brain Project envisages the possibility of complete digital simulation of the human brain. It is illuminating, therefore, to consider the *raison d'être* of their project:

> Understanding the brain is vital, not just to understand the biological mechanisms which give us our thoughts and emotions and which make us human, but for practical reasons. Understanding how the brain processes information can make a fundamental contribution to the development of new computing technology. (EPFL 2020, online)

Whilst the BBP recognizes the 'value' of understanding the brain, since as they put it, 'it gives us' thoughts, emotions and ultimately 'humanity', it appears as an almost secondary benefit. The most important driver in the development of the Blue Brain Project is less about understanding the human mind and more about the development of new more powerful computing technology. So, we have a curious reflexivity at play. By modelling the technology on the structure of the brain, we may come to better understand the functioning of the brain, which in turn helps us to develop new and better computing technology, in order to then model the brain in more detail. It is a perfectly circular argument which depicts the persistent computational metaphor at play in popular understandings of the human brain ever since the beginning of cybernetics.[5] Furthermore, the Blue Brain Project, with its absolute faith in the power of digital

[5] Thanks to the work of mathematician and creator of one of the early computers John Von Neumann and his model of computation, his definition of the essential equivalence of the human brain and the computer continues to influence computing to this day. As Kurzweil (2012) puts it: 'He acknowledges the apparently deep structural differences, but by applying Turing's principle of the equivalence of all computation, Von Neumann envisions a strategy to understands the brain's methods as computation, to re-create those methods, and ultimately to expand its powers' (p. xii). See also the Macy Conferences, set up by Joshua Macy Junior and Frank Fremont Smith in New York between 1946 and 1953.

simulation promises the erasure of the boundary between the functioning of the virtual and the actual and opens up the conceptual question of the possibility of a completely simulated intelligence. But if the brain is nothing more than a highly complex system which we are gaining the ability to reproduce digitally, at what point would the simulation cross over from virtual to real? Or in Hegelian terms, when would it pass from substance to subject?

The Blue Brain Project and other significant developments in the field of brain science (e.g. Neuralink) pose profound questions across many fields of research not just in Artificial Intelligence and engineering but in philosophy, ethics, and even cosmology (Lovelock 2019). In the wake of these groundbreaking developments challenging our notions of biological intelligence, the response from some significant thinkers has been suitably radical. One of them comes from Catherine Malabou. In her (2019) book *Morphing Intelligence: From IQ measurements to Artificial Brains*, Malabou mounts not just a critique of, but a rejection of her previous position on the status of the concept of intelligence as laid out in her (2008) seminal work *What Should We Do with Our Brain?* where she explores the neuro-scientific notion of brain plasticity. In her latest book, Malabou continues what she describes as her examination of the space between biological and symbolic life.

Malabou argues that in light of the development of synaptic chips and the Blue Brain Project, her former thesis on brain plasticity was chauvinistically humanist. The arrival on the scene of the artificial brain signals, in Malabou's view, a watershed moment that changes the status of human intelligence *tout court*. She goes as far as to state that her previous stance on plasticity was wrong, privileging as it did a biologically based model of intelligence. Her new perspective reconceives the brain as something not purely biological with the capacity for plasticity, but rather an entity whose symbolic character does not necessarily require a biological substrate at all.

Malabou lays out a genealogy of the scientific formulation of the concept of intelligence starting in the nineteenth century and tracks its appropriation by historians, psychologists, biologists and philosophers and the disputes that arose from their discussion over its meaning. Malabou goes on to outline what she calls the three main metamorphoses of intelligence, which she calls firstly; genetic fate; secondly epigenesis and synaptic simulation, and lastly the power of automatism. The first metamorphosis is that pertaining to the invention of the IQ test, and the conceptualization of intelligence as a measurable entity for the first time in the nineteenth century. This led to the beginning of eugenics from the work of proto-geneticist Francis Galton followed by Alfred Binet and Theodore Simons research in the field of molecular biology and the sequencing of the human genome.

The second metamorphoses is supposedly our current configuration, that which is encompassed by the field of epigenetics, in other words, our generally and widely accepted understanding of intelligence that takes into account the effects of environmental factors, education, culture on the plastic functioning of the brain as the locus of intelligence. This second metamorphoses is largely locatable to the moment of the twentieth century when the relation between history and biology, thanks in part to French epistemologists such as Bachelard, Canguilhem and the latter's students Foucault and Simondon, started to be appreciated in a new light. It is the start of this reappraisal of the 'innatist and preformationist' (Malabou 2019, p. 15) character of intelligence that, Malabou argues, opens the door towards the possibility of the cybernetic brain that came to dominate the late twentieth century giving weight to the computer-brain metaphor.

The third metamorphoses is still to come. This would entail the removal of the boundary between nature and artifice; a conceptualization of intelligence that allowed for the simulation of brain activity to constitute something more complex and significant than a biopolitical 'threat to humanity'. Rather, it would signal a next stage in the evolution of the concept of intelligent life itself. On this score, we would have to approach Artificial Intelligence in a completely different way, and ask: what do we really mean when we call intelligence 'artificial'? This is in part what Malabou's book attempts to address; the unsaid and undisclosed dimensions of the concept of intelligence that lead us into prejudice as to what *counts* as intelligent life. She asks:

> how, then, should we situate artificial life in relation to biological and symbolic life? Is it an intruder, ever foreign and heterogeneous to them both,

> existing only as a threatening replica? Or is it, rather, the necessary intermediary that enables their dialectical interrelation? (p. xvii)

Indeed, how should we situate artificial life? Malabou does not attempt to resolve this question in the book, and the question of life in relation to intelligence remains indeterminate. In the postscript to *Morphing Intelligence*, Malabou admits that while writing the book, she had not yet realised 'the extent to which the issue of intelligence, particularly Artificial Intelligence, had become a pressing issue, one bound up with significant social, political, legal and economic implications' (p. 145). It was the enthusiastic reception of her book that made her aware that society once again expresses a 'deep and urgent *need for philosophy*' (p. 145; *italics in original*); a need for new tools to help it address the urgent questions posed by AI as a 'transformational technology' that, by virtue of its challenge to the traditional structures of information systems brings about 'a total upheaval of being-in-the-world' (p. 146). The nature of these challenges, Malabou argues, require a philosophical approach that would allow us to apprehend:

> rationally and without delusion [...] a radical revolution not only in the conditions of thought, knowledge and expertise—notions commonly associated with intelligence—but in every field of activity, affectivity and the human psyche. (pp. 145–6)

This point is underscored by Malabou's parting shot, a quote from Freud's (1917) *Introductory Lectures on Psychoanalysis* on the third blow to human narcissism that was psychoanalysis.[6] Reflecting on the transformational nature of AI, Malabou (2019) asserts that the "fourth blow", as she sees it, will be the 'capturing of intelligence by its own simulation, exceeding and transcending it' (p. 164).[7] After reflecting on the upheaval that is AI, Malabou posits that:

---

[6] The first and second being the Copernican cosmological blow and second Darwinian evolutionary blow.

[7] It must be noted however that a fourth blow to human narcissism had previously been characterized by Donna Haraway (2008) as the 'informatic or cyborgian wound' in *When Species Meet.*

> the challenge is to invent a community with machines together, even when we share nothing in common with them. Never will there be a community of machines. The automatic creation they are capable of will have a political platform and ethical texture only if we endow them with it. (p. 161)

It is this final summation of her position that seems paradoxically at odds with the spirit of Malabou's book. For why, given AI's status here as the fourth blow to humanity, would machines not be capable of politics, an ethical act or the founding a community? It seems that whilst Malabou rightly refrains from reifying intelligence she falls into precisely the trap she is warning others of: viewing Artificial Intelligence as some perfectly spherical and discrete Other, unknowable and absolute.

It is interesting then that it is the concept of *stupidity* which Malabou notes, becomes absolutely pivotal in the tradition of French thinking on intelligence. Significantly, as she points out, the genealogy of modern French thought on the question of intelligence from Proust to Flaubert to Valery has always been accompanied by a reflection on stupidity, to which Deleuze's *Difference and Repetition* and Derrida's *The Beast and the Sovereign* would subsequently devote much thought, making stupidity the 'object of a properly transcendental question' (Deleuze cited by Malabou 2019, p. 7). As she puts it: 'a single word, "intelligence," characterizes both genius—natural intelligence—and machines—Artificial Intelligence. A gift is like a motor: it works by itself and does not come of itself. In this sense, then, it is stupid' (p. 8). Malabou concludes that intelligence and stupidity are but one, qualifying their respective fruitless striving for an essence as nothing more than 'ontological stubbornness' (p. 54).

> If such is the case, then why not give up intelligence as an independent philosophical question? The defenses are finally gathering to form this position: ultimately, the ontological void of intelligence is never as evident as in the stupidity of ontology. A stupidity that is, perhaps, not so very distinct from the stupidity of psychology. (p. 54)

Despite her emphasis here on the question of stupidity, Malabou maintains her commitment to rationality as key to a better understanding of the possibilities of artificial forms of intelligence. It is curious then that Malabou, given her long standing dialogue with Freudian and Lacanian psychoanalysis,[8] raises the question of stupidity, given its relationship to enjoyment. For stupidity appears at various times in Lacan's seminar. In fact, he (1998) says plainly that '[t]he signifier is stupid' (p. 20) and more poetically: 'if an angel has such a stupid smile, that is because it is up to its ears in the supreme signifier. To find itself on dry land would do it some good—perhaps it wouldn't smile anymore' (p. 20). Stupidity, whilst not systematized by Lacan in theoretical terms is nevertheless emblematic of much of his conceptual endeavors. His (1988) three passions, love, hate and ignorance could all be called *stupid*. The latter passion, ignorance is perhaps the most self-evidently stupid, of which Lacan (2018) speaks in *Seminar XIX* in relation to the savoir of the analyst; ignorance being the basis of the analytic relationship; a level of knowledge attributed to the analyst, which in fact resides in a blind spot in the analysand themselves.

Then in *Seminar XX* ignorance becomes a divine attribute. Lacan (1998) reminds us that 'Freud arms himself with Empedocles statement that God must be the most ignorant of all beings, since he does not know hatred' (p. 91). To this ignorant and impotent Other, which forms the basis for both love and hate, Lacan gives pride of place in his graphs of sexuation as the barred Other. Formally speaking the three passions love, hate and ignorance as the three primordial forms of enjoyment are all stupid. I would argue that stupidity is precisely the blind spot necessary for 'real' intelligence. Not because of some mysterious attribute of "soul" or "consciousness" but because of the strictly formal operation of stupidity constitutive of enjoyment. The relationship between intelligence and stupidity is therefore not a simple question of opposites. Stupidity is precisely the part of intelligence that intelligence itself cannot see, the place from which the subject of intelligence *enjoys* itself.

While Malabou admirably revises what she perceives as her previous conceptual errors regarding the possibilities of machine intelligence, our

[8] See for example Adrian Johnston and Catherine Malabou (2013) *Self and Emotional Life: Philosophy, Psychoanalysis and Neuroscience,* in which the two authors engage in a debate on the potential challenges that the materialist turn in continental philosophy poses for psychoanalytic accounts of the subject.

conceptions of thinking and its relationship to the brain, she fails to consider the irreducible relationship between intelligence and its *constitutive failure*, which enters her discourse under the rubric of *stupidity*. In this sense, it is not the failure to conceive of a 'community of machines' that makes Malabou fall back on the kind of human exceptionalism that she attempted to escape in the first place, but the idea that intelligence can, by means of a careful process of reverse engineering, be eclipsed by its own simulation. This points us not to another poorly disguised argument for human exceptionalism (i.e. 'not all of the subject can be captured by simulation' or 'there is always a remainder that escapes simulation'), but a concern with the failure inherent to intelligence as such. A failure, furthermore, that recuperates the dimension of enjoyment which Malabou omits from her epistemology and which I see as vital to the challenge that Malabou sets us.

Malabou's opening gambit is to critique her previously held position on the question of human intelligence in light of new discoveries about the brain and new technologies, which put the unique status of the human brain under erasure. However, her acceptance of the genuinely philosophical problem of the synthetic brain still evades the central problem of subjectivity, which resides in the structure of enjoyment, or in psychoanalytic terms the sexual non-relation.

Malabou begins the conclusion to her investigation into the metamorphoses of intelligence with the following statement:

> IN THE END, intelligence is not ours, and it's not theirs either. This resistance to appropriation derives from the ontological paradox that constitutes it: intelligence has no being and cannot, therefore, belong to anyone. (p. 139)

By 'theirs' she is referring to those who wish to instrumentalize and appropriate the wholesale concept of intelligence in the synthetic recreation and simulation of the human brain (i.e. the neuroscientists of the Blue Brain Project). So Malabou admits intelligence has no *being*. In a sense an archetypal Lacanian position. Being is lost by virtue of existence, that is to say as soon as intelligence is reified and instrumentalised it ceases to be intelligent and becomes stupid. Yet paradoxically it is precisely this stupidity which gives rise to the possibility of the conceptualization of intelligence itself. Intelligence is that which is always trying to coincide with itself and fails; the ontological void, constituted by the dialectic as we shall see between sex and knowledge.

As I will attempt to show towards the final part of this chapter, this third "metamorphosis" in the concept of intelligence that Malabou refers to then is the position which Lacanian psychoanalytic theory arguably already occupies. Intelligence on this score is, as we shall see, from the outset a paradoxical category at once artificial and stupid.

2.4 Anti-Philosophy: Thinking or Being?

Following Malabou's characterization of the stupidity of ontology; the enjoying blind spot of intelligence; Pasquinelli's constitutive failure; the originary trauma of rationality, Bratton's gendered and duplicitous AI and Parisi's incomputable (but real) Omega number it seems there is a constellation of crucial psychoanalytic issues emerging. Whilst these thinkers have articulated important and distinct dimensions of the critical treatment of Artificial Intelligence, they all seem to share a common theme; the unspoken omission of the psychoanalytic subject in philosophical and scientific discourses on AI. Lacan describes this problematic thus;

> The philosopher is inscribed [...] in the discourse of the master. This doesn't mean that what he says is foolish; it's even more than usable ... Nor does it mean, mind you, that he knows what he's saying. The court fool has a role to play: being truth's stand-in. He can play it by speaking like a language, just like the unconscious. That he is himself unconscious is of secondary importance; the important thing is that the role should be played. (Cited by Badiou 2018, p. 28)[9]

[9] Cormac Gallagher translates the last part of the final sentence of this quote from Lacan's (2009) *L'Étourdit* as 'the role should be held' (p. 42). Because 'played' in the above translation is more consistent with the relationship between 'role' and 'play', I have chosen not to reference Gallagher's version here, which is the edition I otherwise refer to throughout the remainder of the thesis.

In then to begin articulating how psychoanalysis (as distinct from philosophy) would approach the question of 'thinking' and 'knowledge' and their relationship to AI it is necessary to ground the discussion in the logic of the Lacanian subject. This will ultimately lead us to the underlying concept around which this book revolves; the sexual non-rapport.

Lacan's (2006b) 1966 paper '*Science and Truth*', written especially for the *Cahiers pour l'Analyse* and delivered at the beginning of *Seminar XIII The Object of Psychoanalysis* contains the first textual exposition of Lacan's subject of science, which he equates with the subject of psychoanalysis as that which is represented by 'a signifier for another signifier' (p. 875). Lacan's overarching aim in the paper is to identify the break in modern science heralded by Newtonian physics that leads to the eventual emergence of the subject of science. This involves delineating the passage that starts from Galileo and Descartes in the seventeenth century, 'the century of genius' (p. 857) and culminates in Freud's encounter with the unconscious at the turn of the twentieth century. Fundamentally, the subject for Lacan must be 'rigorously distinguished from the biological individual as from any psychological evolution subsumable under the subject of understanding' (p. 875).

As Ed Pluth (2019) points out, at this stage in his teaching Lacan is trying to work out the status of psychoanalysis as an autonomous discourse, aware of its debt to both structuralism and the natural sciences. He considers the subject to be constituted by a *Spaltung* or splitting; a gap between the subject of the enunciated content and the subject of enunciation, which is encountered in psychoanalytic practice in the form of the symptom. According to Lacan (2006b) psychoanalysis is situated at a specific junction in the history of science, making Freud not an anomaly of the scientific tradition but a direct product of it:

> I am saying, contrary to what has been trumped up about a supposed break on Freud's part with the scientism of his time, that it was this very scientism—which one might designate by its allegiance to the ideals of Brücke, themselves passed down from Helmholtz and Du Bois-Reymond's pact to reduce physiology, and the mental functions considered to be included therein, to the mathematically determined terms of thermodynamics (the latter having attained virtual completion during their

> lifetimes)—that led Freud, as his writings show, to pave the way that shall forever bear his name. (p. 857)

That said, the contribution that Freud makes was in a sense not to scientific knowledge, but to its absence. Psychoanalysis highlights the persistence in the subject of a relation to a 'not-knowing'. It is also in this essay where Lacan makes reference to Descartes' *cogito* as the correlate of modern science, a correlation which Descartes himself had misapprehended given his commitment to the belief in a non-deceitful God. The true significance of the Cartesian cogito is not perceived therefore until Freud, who realised that the real subject revealed in the cogito is not the ego of the 'I think', but the subject of the unconscious who postulates this punctual and vanishing moment of enunciation.

Any attempt to incarnate the subject (i.e. in any biologically determined sense, for example, as 'man') is therefore incompatible with the discoveries of both modern science and psychoanalysis. So fervent is Lacan in his conviction in the futility of pinning down the study of the subject using any of the currently available 'scientific' methods, that he makes reference to psychology as the only discipline deserving of that repugnant appellation "science of man", which he points out has found ways to outlive itself by 'providing services to the technocracy' (p. 730). With reference to Canguilhem's (2016) 1958 essay '*What is Psychology?*' Lacan notes that psychology 'slides like a toboggan from the Pantheon to the Prefecture of Police' (p. 730). Given the biopolitical ring to Lacan's distain for psychology, with its philosophical pretentions to know something universal about man only to then march him down to the police station, one can see how the work of '*Science and Truth*' prefigures Lacan's (2007) elaboration of the four discourses two years later in *Seminar XVII*. The essay contains Lacan's precursor to the four discourses in his elaboration of the three distinct modes of truth—formal, efficient and final—which correspond in their structure to science, magic and religion. These categories of thought, Lacan hints, will in turn find their expression in the differential subjective structures of neurosis, perversion and psychosis and their respective epistemological drives of *Verdrängung*, *Verleugnung* and *Verwerfung*.[10]

Current thinking on the status of 'intelligent life' in the universe from purportedly scientific mindsets, often consists of a baffling mix of magical, religious and scientific thinking, shifting the position of truth as cause within their discourse. Take for example Tegmark's (2017) reflections on the conscious awakening of the universe:

> Before the universe awoke, there was no beauty. This makes our cosmic awakening all the more wonderful and worthy of celebrating: it transformed our universe from a mindless zombie with no self-awareness into a living ecosystem harbouring self-reflection, beauty and hope—and the pursuit of goals meaning and purpose. Had our universe never awoken, then, as far as I'm concerned, it would have been completely pointless, merely a gigantic waste of space. (p. 22)

Tegmark's quasi-religious sentiments of wonder seem to merge seamlessly with his valoration of instrumental rationality and genuine surprise at the sheer miraculousness of it all. Magical thinking, Lacan (2006b) notes, entails a dissimulation of knowledge for the subject of science. This, he says, is one of magic's conditions; it 'involves the truth as cause in its guise as efficient cause' (p. 742). Religious thought, on the other hand, involves an eschatological operation as truth as final cause in the form of revelation. As for science, Lacan says it 'does not want to know anything about the truth as cause'; a form of foreclosure is thus the way in which science must proceed in order to produce knowledge, what he terms a 'successful paranoia' (p. 742). In correspondence with Aristotle's four categories of causation, the missing fourth category would then be material causality, which subsequently will come to be occupied by the discourse of psychoanalysis in *Seminar XVII*. All four relations to truth will later become

[10] As Pluth (2019) remarks in his commentary on the essay:

> [Lacan's] remarks about magic and religion earlier can be thought of in terms of repression and negation respectively. While he was explicit about religion involving a negation of the truth as cause, the connection between magic and repression was never made clear. Perhaps his references to the status of knowledge in magic as obscure, could be read of one of the consequences of this repression. Since the truth as cause is repressed in magic—knowledge about magic's efficacy—why it works would remain a mystery to its practitioners and participants. (p. 299)

formalised in a slightly different configuration as the Master's, University, Analysts, and Hysteric's discourse, with science becoming associated with the Hysteric's discourse as that which ceaselessly produces new knowledge.

This formal distinction between truth and knowledge finds its most precise theoretical iteration in the characterization of Lacan's teaching as anti-philosophy, a term which he first used in reference to himself in 1975 in *Peut-être à Vincennes* (Lacan 2001) and subsequently mentioned on only two occasions throughout his seminar to refer to his method. The term however was not invented by Lacan himself. Despite its origins in religious critiques of rationalism, it is crucial to understand that for Lacan, anti-philosophy was not a mysticism nor a religious concern with the ineffable, the human soul or other such metaphysical concepts, rather the point for Lacan was to reassert the primacy of formalization, and the structure of the production of knowledge as distinct from philosophical truth.

In terms of the psychoanalytic approach to AI, anti-philosophy offers us an alternative perspective on the stakes of the debate on AI than we have thus far encountered. It is not so much a question of finding 'new paradigms' to think AI but rather of elucidating exactly how the Lacanian subject is already present in our conception of Artificial Intelligence, and what this mean for the relationship between the two. These inherent 'impossibilities' that we have identified reside at the very kernel of both psychoanalysis and AI.

Whilst anti-philosophy originated in the eighteenth century as a reaction to French enlightenment thought, by critics who sought to defend church authority and religious dogma against the rationalist tide that was descending over Europe, for Lacan, as Samo Tomšič (2016) notes, the term took on a completely different, not to say opposite meaning. Anti-philosophy in this new post-Freudian context was on the contrary linked by Lacan to the modern scientific revolution and its ramifications for the pre-modern Aristotelian orientation in philosophy and science. For Lacan, the term anti-philosophy was not a mere superficial rejection of philosophy, but rather an interrogation of its formal structure; the discursive position of truth in relation to knowledge, to be precise it is an interrogation of the 'imbecilities of philosophy' (Lacan 2001) in Lacan's view (the stupidity of enjoyment). As Tomšič (2016) notes, antiphilosophy was to form one of the pillars of Lacan's *quadrivium* in his proposition for the transition of psychoanalytic knowledge, the other three disciplines being linguistics, mathematical logic and topology. According to Tomšič these other three disciplines are dependent upon:

> three crucial decentralizations conditioned by scientific modernity: decentralisation of language which suspends the organonic (pragmatic) theory of language; decentralization of knowledge, which detached it from the human observer; and finally decentralization of space which progressively gave rise to non-Euclidian geometries and restructured the space of thinking. (p. 102)

These three domains, Tomšič concludes, linguistics, mathematics and topology each could be said to be sciences uniquely concerned with specific dimensions of the Lacanian organon, symbolic, real and imaginary respectively. The fourth term in this quadrivium anti-philosophy, therefore, would function as the *sinthome* tying together the borromean trio as the discipline which resumes the fundamental lesson of psychoanalysis the 'decentralization of thinking' (ibid.).

Lacan's consistent engagement with the work of important philosophical figures was integral to the different stages of development of his psychoanalytic theory. But it was Lacan's endeavor to mathematize the transmission of psychoanalysis, using a plethora of symbols, letters and algebraic formulas and diagrams, which was a direct attempt to escape the bastardisation and misappropriation, which befell the Freudian discovery given its deceptive appearance as "ordinary language" open to hermeneutic abuse. The matheme therefore was, in the Greek sense of '*ta mathemata*', that which can be transmitted without loss (Johnston 2014, p. 254). One of the key agendas of Lacan's anti-philosophy according to Adrian Johnston is:

> [T]o draw attention to the fact that the philosophical (and quotidian) conception of psychoanalysis as a depth hermeneutics in search of the profound meaning of psychical suffering is a hopelessly wrongheaded misreading of Freud and his place in the history of ideas. (p. 255)

And as Johnston notes, it is thanks to the scientific rupture signified by the name Galileo, that according to Lacan the conditions for Freud's discovery became possible in the first place. Superficial rebuttals of the claim to call Lacan an anti-philosopher then often rely on an objection that what Lacan meant by philosophy was either a proxy for university discourse (as he elaborated in his four discourses in *Seminar XVII*), or that on the other hand the philosophy he had in mind was that of the Deleuzo-Guattarian kind, which was ostensibly mounting its anti-Oedipal attack on psychoanalysis in the post '68 revolutionary era. These strands of straw man argumentation, however, do not get to the crux of what is really at stake in Lacan's significance in relation to the discipline of philosophy. Alain Badiou, who devoted an entire seminar series to Lacan's anti-philosophical endeavors (2018) designates him anti-philosopher exemplar along with figures such as Gorgias, Pascal, Rousseau, Kierkegaard, Nietzsche, and Wittgenstein before him, for Badiou modern philosophy is only possible after first traversing Lacan. Indeed, he places Lacan as third in a succession of such pivotal figures in the history of philosophy as Plato and Kant.

According to Matthew Sharpe's (2015) reading, Lacan's claim to be an anti-philosopher in his later teachings has a number of related origins. Firstly, the concern with philosophy as a historical cultural practice of extracting knowledge. This Lacan derives from the influence of Alexandre Kojève and his reading of Hegel's Master-Slave dialectic. On this account, the master was never the man of knowledge, but rather the warrior prepared to fight to the death in the struggle for recognition. The master converts the slave's practical knowledge or *technai* (his lived *savoir faire*) into theoretical knowledge (*epistēmē*). Sharpe points us here to the primal scene of this paradigm of philosophy, which Lacan (2007) locates in *Seminar XVII* as Plato's *Meno*, in which the slave boy is made to recall mathematical knowledge that he didn't know he had through Socrates questioning: 'refer to the *Meno*, where it is a question of the square root of 2 and its incommensurable. There's someone who says. "Hey look, get the slave to come over, that little fellow can't you see he knows"' (Lacan 2007, p. 22). For Lacan it is this transubstantiation of knowledge from technical know-how, (the slave's drawings in the sand) into theoretical episteme (i.e. Euclidean geometry) that constitutes pre-modern philosophy. 'Philosophy in its historical function is this extraction, I would almost say this betrayal, of the slave's knowledge, in order to obtain its transmutation into the master's knowledge' (ibid.). Whilst this characterization of philosophy as the Master's discourse, is susceptible to the critique that it homogenizes the whole of philosophy into one type of thinking, or one disciplinary form, Lacan will later shift his anti-philosophical critique to a slightly different target. In *Seminar XX*, Sharpe (2015) shows how Lacan mounts his critique again, this time for the types of metaphysical claims philosophers *per se* are prone to making. Again, Lacan (1998) reaches back into the history of philosophy, this time to the pre-Socratics. First of all, it can be said that we have changed the thinking subject considerably:

> Since, the "I am thinking" that presupposes itself, grounds existence, we have had to take a step, that of the unconscious … The subject is not the one who thinks. The subject is precisely the one we encourage, not to say it all (*tout dire*), as we tell him in order to charm him—one cannot say it all—but rather utter stupidities. That is the key. (p. 22)

Here Lacan is referring to what he sees as the fundamental presupposition of pre-modern philosophy deriving from Parmenides of Elea's poem, specifically this passage:

> For without the being in relation to which it is uttered you cannot find thinking. For there neither is nor shall be anything outside of being, since fate (Moira) bound it to be whole and immovable. For that reason, all these will be mere names which mortals have laid down, convinced that they were true: coming-to be as well as passing away, Being as well as non-Being, and also change of place and change of shining color. (Cited by Sharpe 2015, p. 11)

In response to this, Lacan (1998) notes that:

> It is precisely because he was a poet that Parmenides says what he has to say to us in the least stupid of manners. Otherwise, the idea that being is and non-being is not, I don't know what that means to you, but personally, I find that stupid. (p. 22)

The elision of thinking and being, Lacan takes as paradigmatic of the whole of Western philosophy right up through to Hegel's 'the real is the rational' (Sharpe 2015, pp. 11–12). According to Lacan it is an error to assume that all that can be thought necessarily can also be or that the world is a mirror of the subject who thinks it. As Sharpe (2015) points out, this imaginary gestalt of the oneness of being and thinking Lacan ascribes to some fundamental pre-modern axioms that derive from the Aristotelian notion of the spherical nature of the universe. He sees the decisive break with the advent of modern science taking his cue however, not from the Copernican revolution that displaces the earth from the center of the universe, but rather from the Keplerian discovery of the elliptical orbit of stars. Displacing all the celestial bodies instead to one side of a bipolar ellipsis, the other point always being an empty space.

Ultimately, Lacan sees the beginning of the subject of modern science as Descartes' cogito. Lacan (2006b) regards the Cartesian cogito not as hypostatized being that thinks but a punctual evanescent moment. This performative character of the cogito guarantees only that there is thinking, not that there is a thing that thinks. As he remarks in 'Science and Truth':

> Which is why it is worth restating that in the test of writing *I am thinking: "therefore I am"*, with quotes around the second clause, it is legible that thought only ground being by knotting itself in speech where every operation goes right to the essence of language. (p. 734)

In *Seminar XI* Lacan gives his account of the cogito 'as a forced choice between cogito and sum' (Dolar 1998, p. 18). Here Lacan makes the distinction between thinking and being; one must choose between the two. If one chooses thinking one must give up being and vice versa. Lacan's point, in this forced choice is that, sum doesn't follow once one has made the first step. Thought depends on the signifier, which turns the subject into the empty point of enunciation, instead of founding his/her being. In the place of the supposed certainty of the subject's being, there is just a void. As Dolar explains: 'It is not the same subject that thinks and that is; the one that is, is not the one that thinks, even more the one that is, is ultimately not a subject at all' (p. 19). Lacan's (1977) point, simply put, is that being and thinking are different concepts. Returning to our earlier philosophical approaches to Artificial Intelligence, their concerns with failure, dupery, incomputability and stupidity we may recognize a certain impossibility emerging. This impossibility arises by virtue of what Lacan would call the split or barred subject. It is precisely this split between being and thinking that radically changes the way we may conceptualize AI as a 'thinking thing'. This split between being and thinking at the heart of Lacan's theory of the subject, casts a different light on philosophical conceptualizations of AI as a discrete form of intelligence in contradistinction to so called 'real intelligence'. The subject of knowledge is already one riven between the truth of the enunciation and the enjoyment produced discursively. Lacan's (1998) main contention against philosophy then was its claim to sense, which for him was what differed about the *truth* of psychoanalysis, since 'the unconscious is the fact that being, by speaking, enjoys' (pp. 118–119). Perhaps AI is best conceived not as a thing which thinks, but a thing which is thought, or as Lacan (1998) puts it:

> The problem with the kind of science I qualify as traditional, because it comes to us from Aristotle's thought, is that it implies that what is thought of (*le pensé*) is in the image of thought, in other words that being thinks. (p. 105)

The human subject is *spoken by* language. It is because of this enjoyment, that the truth of psychoanalysis, as he (2009) articulates in *L'Étourdit* is an *ab-sens*; a lack of sense. This ab-sens is the hole in symbolization that is otherwise referred to as the *non-existent sexual relation*. The Lacanian sexual non-relation is ultimately located within the dimension of language as the organ of castration, and thus the creation of jouissance. The philosophical question of truth is thereby supplanted by the *anti-philosophical* psychoanalytic question of enjoyment. This means that metaphysical knowledge becomes sexual knowledge. It is the question of enjoyment, in lieu of 'sense' that is the pursuit of the psychanalysis (as opposed to the philosophy) of artificial Intelligence. In the following chapter I will examine the *object* of this enjoyment.

Bibliography

Badiou, A. (2008) *Number and Numbers*. Cambridge: Polity Press.

Badiou, A. (2018) *Lacan: Anti-Philosophy 3*. New York: Columbia University Press.

Bostrom, N. (2014) *Superintelligence: Paths, Dangers, Strategies*. Oxford: Oxford University Press.

Bratton, B. (2015) 'Outing Artificial Intelligence: Reckoning with the Turing Test' in M. Pasquinelli (Ed.), *Augmented Intelligence and its Traumas*, pp. 69–80. Lüneburg: Meson Press.

Cahiers Kingston (2012) Synopsis of Jacques-Alain Miller, 'La Suture: Elements de la logique du signifiant'. Available (01.03.20) at: http://cahiers.kingston.ac.uk/synopses/syn1.3.html.

Canguilhem, G. (2016) What is Psychology? *Foucault Studies* 21: pp. 200–213.

Dolar, M. (1998) 'Cogito as the Subject of the Unconscious' in S. Žižek (Ed.), *Cogito and the Unconscious*, pp. 11–40. Durham: Duke University Press.

EPFL (2020) Blue Brain Project. Available (01.03.20) at: https://www.epfl.ch/research/domains/bluebrain/.

Freud, S. (1917) 'Introductory Lectures on Psycho-Analysis ' in (2001) *The Standard Edition of the Complete Psychological Works of Sigmund Freud, Volume XVI (1916–1917): Introductory Lectures on Psycho-Analysis (Part III)*, pp. 243–496. London: Vintage.

Hammond, K. (2018) TedX: A New Philosophy on Artificial Intelligence [Video File]. Available (01.03.20) at: https://www.youtube.com/watch?v=tr9oe2TZiJw&t=179s.

Haraway, D. (2008) *When Species Meet*. Minneapolis: University of Minnesota Press.

Johnston, A. (2014) *Adventures in Transcendental Materialism: Dialogues with Contemporary Thinkers*. Edinburgh: Edinburgh University Press.

Johnston, A., & Malabou, C. (2013) *Self and Emotional Life: Philosophy, Psychoanalysis and Neuroscience*. New York: Columbia University Press.

Kurzweil, R. (2012) Foreword. In J. Neumann (Ed.), *The Computer and the Brain*. New Haven: Yale University Press.

Lacan, J. (1977) *The Seminar of Jacques Lacan Book XI: The Four Fundamental Concepts of Psychoanalysis*. London: Karnac Books.

Lacan, J. (1988) *The Seminar of Jacques Lacan Book I: Freud's Papers on Technique 1953–1954*. London: W.W. Norton & Company.

Lacan, J. (1998) *The Seminar of Jacques Lacan Book XX: Encore—On Feminine Sexuality, the Limits of Love and Knowledge 1972–1973*. London: W.W. Norton & Company.

Lacan, J. (2001) 'Peut-être à Vincennes…' in J. Lacan (Ed.), *Autres écrits*, pp. 313–315. Paris: Éditions du Seuil.

Lacan, J. (2006a) 'Logical Time and the Assertion of Anticipated Certainty' in J. Lacan (Ed.), *Écrits*, pp. 161–175. London: W.W. Norton & Company.

Lacan, J. (2006b) 'Science and Truth' in *Écrits*, pp. 726–745. London: W.W. Norton & Company.

Lacan, J. (2007) *The Seminar of Jacques Lacan Book XVII: The Other Side of Psychoanalysis*. London: W.W. Norton & Company.

Lacan, J. (2009) L'Étourdit. *The Letter (Irish Journal for Lacanian Psychoanalysis)* 41: pp. 31–80.

Lacan, J. (2016) *The Seminar of Jacques Lacan Book XXIII: The sinthome*. Cambridge: Polity Press.

Lacan, J. (2018) *The Seminar of Jacques Lacan Book XIX: … Or Worse*. Cambridge: Polity Press.

Liu, L.H. (2010) *The Freudian Robot: Digital Media and the Future of the Unconscious*. Chicago: The University of Chicago Press.

Lovelock, J. (2019) *Novacene: The Coming Age of Hyperintelligene*. London: Penguin.

Malabou, C. (2008) *What Should We Do with Our Brain?* New York: Fordham University Press.

Malabou, C. (2019) *Morphing Intelligence: From IQ Measurement to Artificial Brains*. New York: Columbia University Press.

Miller, J-A. (2012) 'Suture (Elements of the Logic of the Signifier)' in P. Hallward & K. Peden (Eds.), *Concept and Form Volume One: Key Texts from the Cahiers pour l'Analyse*, pp. 91–101. London: Verso.

Parisi, L. (2015) 'Instrumental Reason, Algorithmic Capitalism and the Incomputable' in M. Pasquinelli (Ed.), *Alleys of Your Mind: Augmented Intelligence and its Traumas*, pp. 125–137. Lüneburg: Meson Press.

Pasquinelli, M. (2015) 'Introduction' in M. Pasquinelli (Ed.), *Alleys of Your Mind: Augmented Intelligence and its Traumas*, pp. 7–18. Lüneburg: Meson Press.

Pluth, E. (2019) 'Science and Truth' in S. Vanheule, D. Hook & C. Neill (Eds.), *Reading Lacan's Écrits: From 'Signification of the Phallus' to 'Metaphor of the Subject'*, pp. 268–307. London: Routledge.

Searle, J.R. (1980) Minds, Brains, and Programs. *Behavioral and Brain Sciences* 3(3): pp. 417–457.
Sharpe, M. (2015) Killing the Father, Parminedes: On Lacan's Antiphilosophy. *Continental Philosophy Review* 52: pp. 51–74.
Tegmark, M. (2017) *Life 3.0: Being Human in the Age of Artificial Intelligence*. London: Penguin.
Tomšič, S. (2016) 'Psychoanalysis and Antiphilosophy: The Case of Jacques Lacan' in A. Cerda-Rueda (Ed.), *Sex and Nothing: Bridges from Psychoanalysis to Philosophy*, pp. 81–103. London: Karnac Books.
Von Neumann, J. (2012) *The Computer and the Brain*. New Haven: Yale University Press.
Wright, C. (2018) 'Lacan's Cybernetic Theory of Causality: Repetition and the Unconscious in Duncan Jones' *Source Code*' in S. Matviyenko & J. Roof (Eds.), *Lacan and the Posthuman*, pp. 67–88. Basingstoke: Palgrave Macmillan.

Filmography

Cameron, J. (Director) (1984) The Terminator [Film]. US: Orion Pictures.
Columbus, C. (Director) (1999) Bicentennial Man [Film]. US: Buena Vista Pictures.
Spielberg, S. (Director) (2001) A.I. Artificial Intelligence [Film]. US: Warner Bros. Pictures.

3

The Artificial Object

The myth of the wonder-working functionality of the world *is correlated with the phantasy of the wonder-working functionality of the* body. *There is a direct link between the paradigm of technical action executed by the world and the paradigm of sexual action executed by the subject; and in this perspective the gizmo, the ultimate tool, is basically the substitute for the phallus, the operative medium of function* par excellence.
—Jean Baudrillard (2005b, p. 126)

3.1 Black Mirror and the Lathouse[1]

In 2016 Elon Musk launched Neuralink, a company dedicated to the creation of a "brain-computer interface" which could be implanted into the human brain painlessly, safely and efficiently. According to Musk, the implant, consisting of thousands of microfibers, which could penetrate the brain tissue via robotic surgery in order to mimic the firing of

[1] Elements of this section have appeared in adapted form. See Millar, I. (2018) Black Mirror: From Lacan's Lathouse to Miller's Speaking Body. *Psychoanalytische Perspectieven 36*(2): pp. 187–204.

I. Millar, *The Psychoanalysis of Artificial Intelligence*, The Palgrave Lacan Series,
https://doi.org/10.1007/978-3-030-67981-1_3

neurons, could potentially be used for the treatment of brain injuries including paralysis or blindness. The interface has so far only been tested on rats and monkeys, but Neuralink are currently in the process of applying for FDA permission for the first human test subject. Fundamentally the aim of Musk's company is to merge AI with human brains, creating a superhuman form of intelligence, yet one that humans can potentially still be "in control" of. Strangely enough, we have as yet no data about the experience of a human subject who has had the interface implanted.

The anxieties provoked by the potentials of this kind of technology are explored in Charlie Brooker's dystopian science fiction drama series *Black Mirror*. The series illuminates ethical and conceptual problems raised by new forms of AI which breach the body and brain. Technologies that fundamentally reorganise our libidinal economies, systems of knowledge production and perception of reality.[2] Through placing the human subject in scenarios where the real, symbolic and imaginary dimensions of experience are displaced, *Black Mirror* depicts the anxieties of the human as it becomes indistinguishable from Artificial Intelligence.

The episode *Arkangel* (Foster 2017) follows the story of a mother Marie and her daughter Sarah. Marie who, having been terrified by the experience of losing sight of her little girl one day and discovering her down by a rail road track, decides to take part in a scientific trial which places a permanent monitoring and tracking system into the child's brain. The implant is connected to a device allowing Marie to have complete surveillance over Sarah's whereabouts, and her bodily vitals, including monitoring of cortisol levels to enable instantaneous intervention if she witnesses a distressing scene. The device can then filter out the image or audio that the child receives. It not only provides geographical, and physiological data but allows Marie to have the audio-visual information that her child is experiencing at any moment. Sarah becomes completely subject to her mother's desire.

---

[2] Not surprisingly the series has attracted much attention from cultural critics and philosophical speculation. 2019 saw the publication of an anthology of essays entitled *Black Mirror and Philosophy* (Johnson 2020), in which each episode of the five series was analysed by a different author with the view to tackling a specific philosophical question. What is surprising however, given the obvious psychoanalytic scope of the series, is that the index of references contains only one entry on Freud (one paragraph on the death drive).

As the little girl grows up, Marie realises that her over-protective measures may have had detrimental effects. Frustrated by her own inability to understand the faces of people in distress or see what happens when someone is injured, we witness the now 8 year old Sarah pricking her finger furiously with a pencil only to see the blood trickling out becomes blurred by her monitoring system. Marie finds her distressed daughter stabbing at her hand and decides she must take her to a therapist. Significantly the mention of autism[3] is made by Marie, thinking her child lacks the ability to understand or read the emotions of others and, by extension, her own. The therapist rejects the hypothesis and suggests that the device may be simply turned off allowing Sarah to start experiencing life unfiltered and un-surveilled by her mother. Marie agrees. It is not long before Sarah is initiated by her school mate into online porn and the grim and violent realities of the world. In the following years, without the gaze of arkangel, she is left to fend for herself, until one night aged fifteen Sarah doesn't come home. Her mother panicking, switches on the arkangel device to find that Sarah is having sex. Marie starts using the device again and witnesses her daughter taking cocaine with her boyfriend.

Her mother continues to monitor her bodily functions and discovers that Sarah is pregnant. Marie drops an emergency contraceptive pill into Sarah's morning smoothie. Sarah throws up in class and finds out soon enough the reason why. Sarah runs home and furiously searches through the house to look for her mother's device. Horrified, Sarah finds that her mother had been monitoring every moment of her life and had not only orchestrated her break up but now had aborted her baby. Marie tries to convince her daughter she was using the device to protect her. But Sarah will not be assuaged and beats her mother over the head repeatedly with her arkangel device. Miraculously, the filtering option has been switched

[3] Whilst the extreme autistic subject may appear to be outside the social bond, given our growing dependence on computational forms of intelligence and the entrance of AI into the social bond, understanding the autistic mode of enjoyment at work is vital. The agenda of the autistic structure to limit thinking to an exquisite use of logic and mathematical systematization betrays a desire to obviate the traumatic entrance of the other's jouissance into the equation. Given the autistic tendency towards particularate systems, is it therefore surprising that the technological and digital revolution has coincided with a massive increase in the appearance and diagnoses of autistic subjects?

back on, so Sarah cannot not see the full extent of the brutality she inflicts on her mother's face.

How may we interpret this dystopian mother/daughter dynamic? As Marie's behaviour becomes ever more controlling with the help of arkangel, the desire of the Mother is completely unfettered, and Sarah becomes the absolute object for Marie. Marie even comes to occupy the obscure extimate position of the gaze from the point of view of her daughter. But this is more than a morality tale of overprotective parenting. Sarah is taken from an almost autistic inability to recognise distress to suddenly having to decode and assimilate an overabundance of enigmatic stimuli and signifiers. More than regulating her desire, the arkangel device operates at the level of the drive. Surely the intrusion of this artificial object has effects 'in the real' for Sarah?

In *Seminar XVII* Lacan (2007) makes reference to the *lathouse*, a device which siphons off enjoyment and indelibly inscribes it in codified form in the *alethosphere* (deriving from alethia—the Greek word for 'truth'). Originally Lacan was referring to the use of tape recorders in his seminar and the way in which they removed and recorded the enjoyment of the voice and codified it into a realm of shared meaning enabling others to also *j-"ouir"* (Lacan (1977) equivocates, *ouir* in French *to hear*) Lacan's voice separately from his body. However, he was not just concerned with the gadgetry involved in this operation, (nor to the commodified form of enjoyment) but more specifically to the possibility of siphoning off bodily enjoyment via some form of apparatus. The implications of his theorizations reached further than to the analogue technology of the day however, and it seems Lacan's allusions were prefiguring the digital revolution which occurred not long after his death. He (2007) noted:

> The world is increasingly populated by lathouses. [...] The lathouse has absolutely no reason to limit its multiplications. What is important is to know what happens when one really enters into a relationship with the lathouse as such. (p. 162)

We can extrapolate that this is what Lacan envisioned to encapsulate the new conditions of the social bond, instigated by the alliance between the discourses of science and of capitalism. He alludes to the idea that the lathouse provides an impossible function:

> It is clear that it is completely impossible to hold the position of the lathouse. [...] It's at the level of the impossible, as you know, that I define what is real. [...] This forms part of the position of the lathouse. (p. 163)

It is remarkable that Lacan here refers to the lathouse as occupying the position of the real, which one could simply equate with the *object a*, yet surely, as a reified product of science, this artificial object has effects *in the real*? Lacan created these neologisms to speculate on the growing importance and implications of these 'little devices' on forms of enjoyment and the sphere of 'formalized truth' of which they are part (pp. 161–162). The alethosphere on the one hand, pre-figures the idea of a spherical, self-contained yet expansive world where the subject is plugged in to her own singular mode of enjoyment yet is recorded by the Other. This we may recognize as describing the state of Baudrillard's 'hyperreal' subjects who exist in a complex system of signs completely detached from any link to so called 'reality'. But today the alethosphere reaches beyond the visual realm of sign value and operates on the infinitely more complex scale of the algorithm.

Lathouses on the other hand, Lacan says are 'the objet a's we find 'at the corner of every street, behind every window [...] designed to be the cause of your desire, insofar as it is now science that governs it' (p. 162). As Lacan explains, the lathouse is a machine, an artificial object for siphoning off enjoyment—a neologism combining the French "vent" for wind, alluding to the breath from the lungs; "venthouse" suction cap; and the Greek word *ousia* for Being. The crucial point Lacan was making was not just that these objects are causes of desire, but that they contain something of the drive. In that sense they are impossible objects that attempt to capture the jouissance of the body which allows the truth of our enjoyment to be recorded by the Other qua alethosphere. Whilst he may not have imagined that we would have a smartphone in our pockets capable of giving us instant access to encyclopaedic information, global news, or even sex with strangers, Lacan was alluding to the suspicion that science would soon have a means of harvesting and registering these objects of desire in such an efficient way as to completely change our way of pursuing them. But in contemporary times it is not only the voice that may be captured by the lathouse, the voice is just one drive object that may be simulated and administered by the work of such a device. The lathouse may be thought of as a function which attempts to drain jouissance from the body, or perhaps more accurately, regulate and administer it.

The arkangel device then is a lathouse, which siphons off Sarah's jouissance but at the same it is plugged in to the alethosphere of truth, registering on the level of the Other. However paradoxically her enjoyment (has been in part) conditioned upon the impossibility of fully understanding the Other's enjoyment. In the denouement of the story, Sarah finds herself unable to cope with the volatility of her situation when, having discovered her mother's monitoring device she is confronted with the *mise en abyme* on her arkangel screen, which shows Sarah an infinite regression of her very own gaze within a gaze. Sarah has been totally alienated from her social interactions, unable to formulate appropriate reactions to the other's suffering nor understand her own, her body becomes maladjusted to the demands of the social bond. Her relationship to her body as foreign to her is made manifest by the arkangel's sovereign position governing her physiological reactions and intervening into the space of fantasy.

If we are to take seriously the notion that technoscience has effects in the real that were previously absent, then surely these objects warrant further speculation than merely as consumer pests? In *Seminar XX* Lacan (1998) states:

> scientific discourse has engendered all sorts of instruments that we must, from our vantage point here, qualify as gadgets. You are now, infinitely more than you think, subjects of instruments that, from the microscope right down to the radiotelevision, are becoming the elements of your existence. You cannot currently even gauge the import of this, but it is nonetheless part of what I am calling scientific discourse insofar as a discourse is what determines a form of social link. (p. 82)

What is the lathouse if not this curious object existing at the limit point between, following Tomšič (2012), scientific discovery and invention?

The question is, apart from dramatizing the potential catastrophe this could provoke, how are we to seriously engage with the conceptual ramifications of the merging of AI with the body in this way? If the standard way of analysing the impact of AI typically resorts to the critique that society is now one completely governed by the instrumentalization of data at the expense of 'human interest', this leaves the conceptual and psychoanalytic potential of the lathouse unexamined.

In a way, we could view Lacan's dual conceptualization of the alethosphere and the lathouse as articulating two distinct modes of relating to the problems of Artificial Intelligence. Firstly, the alethosphere conjures the image of a self-contained world in which meaning circulates according to a system of symbolic coordinates, a machinic order that algorithmically gathers data in the pursuit of an 'acephalic' knowledge. This would be the university discourse in position of agent in Lacan's four discourses (as we shall discuss further in Chap. 4). Furthermore, this would conform to what Lacan, in his critique of 'philosophy' might call 'spherical thinking' as we discussed in Chap. 2. In other words, the mistaken idea that knowledge can be 'absolute'. The idea that all that is necessary to obtain an accurate picture of reality is the collection of enough facts about nature; the fantasy of wholeness that science (in the guise of the university discourse) hungers for. But the lathouse operates on an altogether different level. We could say on the level of the *not-all*. The lathouse is not merely concerned with gathering data, but rather has the potential to intervene into the body at the level of enjoyment, in order to create new effects in the real.

Drawing on Lacan's (2013) short lecture On the Names-of the-Father, it is interesting to note the biblical significance of the title and names of the lead character of this episode. This lecture delivered just before Lacan began his *Seminar XI*, briefly articulates the real, symbolic and imaginary function of the name of the father in the Judeo-Christian tradition and analyses the story of Abraham's command from God to sacrifice his son Isaac. The function of the angel in the biblical story is at the last minute to prevent Abraham from killing his son and instead overseeing the sacrificing of a ram. The meaning of this has been historically and religiously interpreted in many ways. One of the main theological issues is the idea of the authority of the angel over the word of God. Is it a vetoing of the capricious whims of God, or does the angel act as a mere conduit for the message of God to let Abraham know he was being tested? In terms of the Name-of the-Father in the Black Mirror narrative, the Arkangel serves just this ambiguous function. On the one hand it is the conduit of S1 of scientific mastery; a mere messenger of university discourse, and on the other it serves as a putative father, a law giver which acts autonomously, capriciously and without justification, in Lacan's (1992) terms the "Great Fucker". Incidentally, Sarah is also named in Lacan's text. She is the wife of Abraham and the miraculous nonagenarian mother of Isaac, who is already ninety one by the time the sacrificial child was born. The arkangel in this modern parable far from overseeing the safety of Sarah's future child instead facilitates its abortion at the hands of her mother.

The arkangel in the form of lathouse is, in a Freudian sense, a censor but one which changes the very coordinates of sensory experience operating on the level of the generation or indeed the negation of desire and the governance and administration of jouissance. The arkangel device in this instance, instead of heightening reality à la Baudrillard into a hyperreal experience, dampens 'reality', a *hypo-reality*. Less than real.

In relation to the function of the lathouse on the drive body we may here refer to Lacan's (2007) use of the term 'operceive' (p. 160), comprising the words operational and perceive, which seems to sum up the way in which science convenes with object a to create new forms of perceptual capabilities. He writes:

> [t]hat we can find in the place, in the "operceive" in which science comes to be constructed. What I perceive, which I claim to be original, must in effect be replaced by an operceive. Insofar as science only refers to an articulation that only takes form in the signifying order, it is constructed out of something where there was nothing beforehand. This is what is important to grasp if we wish to understand something that has to do with—what?—with the forgetting of this very effect. Being what we are, all of us, to the extent that the field increases by virtue of the fact that science perhaps functions as the master's discourse, we do not know how far- for the reason that we have never known at any point- each one of us is initially determined as small object a. (p. 160)

Here we see the question emerge, if AI in the form of a highly sophisticated lathouse, may survey the desires to such an extent that paternal prohibition is short circuited how does this intervene into the space of fantasy? How does the lathouse act as an administrator of jouissance? This *'opercieve'*—or operception to coin a phrase—seems to allude to the potential for the lathouse to have material effects on the drive body. In other words, do these artificial objects interact with the drive at an *originary* level?

3.2 The Prosthetic God

In his lecture *A Fantasy*, Jacques-Alain Miller (2004) describes what he sees as the new discourse of civilization, where the *object a* is in the place of agent. Miller argues that with the replacement of agricultural civilization with industrial civilization, the so-called compass of nature has been destroyed. Or as he puts it:

> Agricultural civilization finds its bearings through nature, through the invariable cycle of seasons. Of course, there is a history of climate that some well-intentioned people are now reconstituting. But this history changes in no way the invariable cycle of seasons that gave its rhythm to agricultural civilization, so that, in fact, it was possible to find one's bearings and one's symbols in the seasons and the skies. The agricultural real is celestial; it is a friend of nature. With industry, with what has been called the industrial revolution, all that was washed away, little by little. The artifices were multiplied. And now we are forced to notice that the real is devouring nature, that it is being substituted for it and is proliferating. Here we have a second metaphor: the metaphor that substitutes the real for nature. (p. 5)

From here Miller goes on to establish that the new compass of civilization, since its unmooring from the celestial rhythms shown to us so gracefully by nature, must be the *object a*, which is reliable and dependable in that it always maintains its place as a crack in reality. However, there seems to be a paradox at work here. As Žižek (2017) reminds us, there has never been such a thing as nature as far as the unconscious is concerned. He argues, humans never really trusted 'nature' anyway, even doubting if the sun would rise in the morning; hence why they invented so many gods in order to account for the caprice that they experienced at the hands of so-called Mother nature. The real according to Lacan is that which always remains in its place. What Miller seems to be suggesting however, is that this is no longer the case, thanks to the alliance of science and capitalism and the reified objects of technoscience that they produce (Miller 2013a, b). As such, Žižek has accused Miller of a mischaracterization of the real, since in his passing from the real of nature to the "pure lawless real," what goes missing supposedly is the Lacanian real itself as nothing but a pure impasse in formalization, or, in other words, that which cannot be exhausted by the symbolic. So, when Miller (2013a) says:

> Nature was the name of the real when there was no disorder in the real. When nature was the name of the real you could say, as Lacan did, that the real always returns to the same place. [...] I would say that capitalism and science combine, they have combined, to make nature disappear. And what is left by the vanishing of nature, what is left is that which we call the real, that is, a remainder. And, by structure, disordered. The real is touched on all sides by the advances of the binary capitalism-science, in a disordered way, randomly, without being able to recuperate any idea of harmony.

Žižek sees this reification of the real as a form of technophobic naiveté or postmodern obfuscation, but could he be missing the point? Miller goes on to state that the epoch of the real as nature has the function of Other of the Other, whereas our current moment is characterised as the Other *without* Other (Miller 2013b), a reference to Lacan's (2019) conceit in *Seminar VI* that there is 'no Other of the Other' (p. 291). According to Zizek, Miller's view of capitalism and its marriage with technoscience as the 'real outside of the law' is merely a disavowal of the structural antagonism inherent in the real.

To return to the concept of the lathouse as Lacan (2007) put it: 'insofar as science only refers to an articulation that only takes form in the signifying order, it is constructed out of something where there was nothing beforehand' (p. 160). Veronique Voruz (2013) refers to precisely these non-natural objects of technoscience produced by the marriage of science and capitalism in the dimension of the *alethosphere*. She points out that 'in the place of objects of the body, the more classical guises of *object a*, the alethosphere produces lathouses, the objects of the real of science.' The key concern for Miller then in relation to the real, is what is the status of these new objects produced by science? Are they the same as the *objects a* lost from the body either naturally or as effects of the signifier? Miller (2013b) asks: 'the question being are these new objects completely new or are they merely reconstituted forms of primordial *objects a*?' It is Miller's above question that seems key to the relationship between AI and the real of the drive body.

Perhaps to clarify the matter, the question of technoscience and the 'real outside the law' may be addressed with recourse to Freud.

In *Civilization and its Discontents*, Freud (2004) delivers his brief but sceptical vision of modern subjectivity; man as a *Prothesengott* or "prosthetic God". Thanks to the reification of science by technology, the human subject, in a quest to become omnipotent and omniscient, is now endowed with various auxiliary organs. These prosthetic organs however do not fulfil their promise. Man does not become a real god and is constantly chasing new ways to transcend his bodily limitations. Freud remarks that this is testified to by the fact that these auxiliary organs are not one with the organism and can never become such. The prosthetic god is by nature flawed and carries with him an inherent lack. He eternally fails to achieve the fantasy of potency and enlightenment he imagines, and instead employs his auxiliary organs to circulate perpetually around the objects of the drive.

The point here is that Freud is making the distinction between the biological body and the drive body. The living body is split between anatomy and the effects of the signifier. As Tomšič (2012) points out 'Freud's description of the Prosthesengott is not a mere caricature of the subject of technology; it is an alternative description of the subject of the signifier given that the signifier also produces at the level of the body an irreducible non- relation between the biological and the speaking body' (p. 146)—what Lacan will later come to term the parlêtre.

Freud (2004) writes:

> Man has become, so to speak a [prosthetic] god with artificial limbs. He is quite impressive when he puts on all his auxiliary organs, but they have not become part of him and give him a good deal of trouble on occasion … Distant ages will bring with them new and probably unimaginable achievements in this field of civilization and so enhance his godlike nature. But in the interests of our investigations, let us also remember that modern man does not feel happy with his godlike nature. (pp. 36–37)

At the time of Freud's 1930 writing, both civilization and its discontents were already significantly reworked through the emergence of new and radical forms of media (Kittler 1999). But today's invasive biotechnological interventions on the body and brain problematize of the notion of the organ itself. It may seem like Freud is describing the familiar sci-fi anxiety over the future impingement of technology onto the human body but in fact he is saying something more nuanced than it first appears, contrary to how some critics have understood him.

In the series *Technics and Time* Bernard Stiegler (1998) attempts to rethink the relationship between the human and technical objects, or "technics"; a term referring to what he calls organized inorganic matter. Stiegler's project explores a history of technics as *epiphylogenesis*—the preservation in technical objects of epigenetic experience. Epiphylogenesis for Stiegler marks a break with genetic evolution (which, he argues, cannot preserve the lessons of experience), a break which also constitutes the "invention" of the human. As he puts it in *Technics and Time 1*: '[a]s a 'process of exteriorization,' technics is the pursuit of life by means other than life' (p. 17). Stiegler recounts the myth from Plato's *Protagoras* of Prometheus and Epimetheus as the allegory of hominization. Prometheus' forgetful brother Epimetheus takes over his brother's task of allocating qualities to the animal world so as to create the balance of nature. In so doing Epimetheus forgets to leave any qualities for mankind. Having nothing left to give the human, Prometheus steals fire from Hephaestus, the blacksmith of the Gods. Thus, man is bestowed with the *original tool* or technic. From fire man's destiny as technical being is forged. For Stiegler, the human is a creature inherently lacking in qualities or "nature" who props himself up by means of technology or, in other words, *artifice*. Stiegler, drawing on the work of palaeontologist Leroi Gourhan, asserts that all beings before man had two forms of memory: DNA memory and individual memory of the central nervous system. These systems however do not communicate with each other. As such these beings had no way of transmitting individual experience to the next generation. Once man had evolved to bipedal form, freeing his front limbs and displaying his face for communication, the invention of "tools" becomes possible. Tools became the first types of memory support allowing cultural artefacts to be preserved. Thus, technics is just this exteriorization of memory.

Stiegler utilises Plato's dialogues in Meno to explain how the slave boy's shapes drawn in the sand usher in the invention of geometry. With this he emphasises, however, that the Platonic opposition between the intelligible and the sensible—that is, between *logos* and *technê*—makes the material nature of the apparently abstract process of geometrical thinking, impossible to understand. Thus, in the Meno, metaphysics takes shape as *the denegation of the originary technicity of memory*. Plato as the founder of politics is simultaneously the fateful denier of the technicity of thought and, by extension, the technicity of law and political life.

Stiegler's project articulates what he calls the *pharmacological* paradox of our time: the outsourcing of capacities in to the "black box" of digital technologies or *mnemotechnics* (what in Marxist terms is referred to as the proletarianization of human labour) as both a gain and a loss. A gain for technological development and simultaneously a possible loss in human organic capacity; a *becoming-redundant* of the organic in the service of the "artificial". He sees the process of individuation and subjectification as one defined by an articulation of memory. Or what he calls a *grammatization*. This *mnemotechnics* is the exteriorization of knowledge from organic to inorganic matter. Existence or "con-sistence", as Stiegler would say, is then a technical economy of anticipation, memorization or enhancement. Grammatization, Stiegler tells us, is the history of the exteriorization of memory in all its forms; nervous and cerebral memory, linguistic, auditive and visual, bodily, muscular and biogenetic. Once exteriorised, memory becomes the object of social, political and biopolitical controls through the economic investments of social organizations, which reach all psychic organizations by means of mnemnotechnical organs including machine tools, digital gadgets and even household appliances. Paraphrasing Stiegler, delegating operations to machines means the transindividuation processes are altered and, in some cases, dangerously short circuited. Not only are the *analytic* capacities of the understanding delegated to the machines, but the *synthetic* capacities of reason are utterly wiped out. This is why, in Stiegler's view, a thinking of grammatization calls for a 'general organology' (2014). That is, a theory of the articulation of bodily organs, artificial organs and social organs, thereby opening up the positive possibilities presented by the *phamarkon* of technological enhancement and to delimit the toxicity of it.

In his contemporary work, starting with the series *Symbolic Misery*, Stiegler (2014) detects a danger in the pervasive encroachment of advanced forms of technology into the social bond. But rather than the hyperbolic question of superintelligence that Bostrom contemplates as an existential threat at the level of the species, Stiegler is concerned more with the bio-political and aesthetic dimensions at the level of the subject. His marginally less apocalyptic diagnoses for this state of affairs he calls "symbolic misery". In Stiegler's (2014) view, biopolitics has reached such a saturation point whereby a type of symbolic misery is unfolding. In the hyperindustrial epoch there is a complete seizure of the symbolic by technology, audiovisual and digital mechanisms are gradually taking total control over the rhythms of our bodies and "souls". It is for reason that Stiegler (2014) accuses Freud of not quite getting the significance of the technical object in its radical ontological significance. But could it be that Stiegler (2014) misses the point of Freud's use of the prosthetic God, however, when he states that:

> This Prosthetic destiny does not arise in the twentieth century as we might think from a cursory reading of *Civilization and its Discontents*, but represents the originary default of the origin that is the *originary murder* of the father by the weapon that is all technics. And the first of these technics is the knife, that of *Totem and Taboo* just as much as that of the sacrifice of Isaac, but which Freud, not knowing how, was unable to think. (p. 12)

What Stiegler arguably fails to appreciate here is the logic of castration which is precisely the originary murder that Freud (1913) was describing in *Totem and Taboo*. While Steigler accuses Freud of not going far back enough in his theory of the technical object, is it in fact Stiegler who does not go nearly far enough? The knife which kills the totemic father is not in fact the first technics, which indeed Freud knew only too well and is why he had to resort to the medium of myth to express his point. What Freud understood to be the first of all technics, although he perhaps would not have put it that way, was sex. Sexuality, being the original 'artificial' invention that marks the entrance into to the symbolic and the separation from the animal kingdom for the speaking being, and the passage from the biological body to the drive body. Far from predicting a possible time where humans will be replaced by some perfect God-like version of themselves that is all-powerful and all-knowing (in the form of Kurzweil's Singularity for example), Freud recognises the inherent fantasmatic structure of this prosthetic God. Freud is in effect acknowledging the fact that the human *is human* by virtue of his inability to find satisfaction through his auxiliary organs. These organs will only ever circulate hopelessly around their partial drive objects. Much in the same way that Stiegler (1998) hypothesizes the technicity of being as predicated on the founding myth of the fault of Epimetheus, Freud had perceived a certain deficit or fault in the human, (or *hilflosigkeit*) whose existence was sustained by a perpetual prosthesis.

The originary murder staged in *Totem and Taboo* is also the intractable antagonism giving form to the non-existent sexual relation and its resultant masculine and feminine modes of enjoyment as we shall discuss in Chap. 4.

In *What Makes Life Worth Living* Stiegler (2013), returning to the myth of Prometheus, explores the libidinal implications of his position on pharmacology:

> Fire is the *pharmakon* par excellence. As civilizing process, it is constantly at risk of setting fire to civilization. As the emblem common to technics and desire, it constitutes and articulates a *dual logic of the necessary default:*
>
> - that logic shown by Freud to operate via the 'perfecting of organs', that is, as technics, a process that interminably *displaces* organic and organological default, that default that is necessary; and
> - that logic that Lacan attempted to describe 'lack—which is *precisely not a mere lack*, but on the contrary *necessary*: the stoic quasi-cause.

> Technics, which thus pharmacologically constitutes the default insofar as it forms the horizon of desire, simultaneously opens two antagonistic yet inseparable paths: that of the drives, and that of sublimation. (p. 24)

Thus, as Stiegler explains, we cannot understand the pharmacology of technics without taking recourse to the libidinal economy of the subject. How then would we address Stiegler's pharmacological question in relation to the '*operceptual' capacity* of AI? And does this get to the crux of Žižek and Miller's disagreement over the so called '21st century real'? How does the question of the subject and the drive figure in this dialectic of AI and the human body? Stiegler's allusion to the changing relationship that the *drive* has with its technological objects has resonances with the concept of the lathouse, as an extimate 'non-natural' object that administers the jouissance of the body.

In Stieglerian vein, the mnemotechnological potential of the drives to be exteriorised, augmented and codified into the alethosphere calls into question the dimensions of jouissance and its relationship to knowledge. A grammatization (as Stiegler might call it) of the techno-organological nature of the symptom is what is at stake. It is in this context where the lathouse takes on significance as a different form of object. Based on this reading if *object a* is on the side of the metonymy of desire, then the lathouse is on the side of the drive and the body (Fig. 3.1).

This corresponds to the shift from the metonymic movement of desire in the pursuit of 'meaning' to the circulation of the drive around the ab-sex-sense of jouissance. But what exactly does this mean? We can say that the lathouse is an extimate object which converts interiority into exteriority and vice versa. The lathouse it must be emphasized is therefore an *object* split between the biological body and the drive body. If we recall in the case of Roko's Basilisk, with the singulatarian (and phallic) speculation on the merging of the human and AI, the question of the body is either totally absent, or is a site of perpetual torment, but this is precisely

Object a = Desire/Meaning : Lathouse = Drive/Sense-Absexe

Fig. 3.1 Object a and the lathouse

the fantasy of wholeness so prevalent in discourses on AI that the lathouse as *not-all* evades.

In *What is Sex?* Alenka Zupančič (2017) draws attention to Andre Platonov's *The Anti-Sexus.* Written in 1926, the work is a fictional brochure and "translated" from French by Platonov which:

> advertised an electromagnetic instrument promising to relive sexual urges in an efficient and hygienic manner. The device available in both male and female models, had a special regulator for the duration of pleasure and could be fitted for either personal or collective use… the company's mission [is] to "abolish the sexual savagery of mankind"…The Anti-Sexus…has many benefits and applications: it is perfect for maintaining soldiers' morale during wartime, and for improving the efficiency of factory workers…It also fosters true friendship and human understanding by taking sexual folly out of the equation. (Schuster cited by Zupančič, p. 27)

Zupančič asks what we are to make of the logic behind the strange gadget which promised to emancipate mankind from its torturous shacking to the pursuit of sexual relations. She identifies the two suppositions behind the creation of the Anti-Sexus device as follows:

> Sexuality is problematic because it involves the Other who is unpredictable, unreliable (has her own will, caprices, indispositions…) or simply unavailable. On the other hand…our relations with others are complicated and conflict ridden because expectations and demands concerning sex are always in the air…sex stands in the ways of good social relations. (p. 27)

So the double quandary here is on the one hand the anti-sexus promises to remove sex from the Other and by the same token remove the Other from sex, as a result of which we get two separate entities; on the one hand a sexless other who we may relate to in a purely "spiritual" way, purged of any debasing sexual elements and on the other hand we get a pure sexual substance, which we can enjoy at any time without the need for the other's presence, or more radically, without the need for their existence. Here Zupančič identifies the contradiction at work in this imaginary scenario, which sums up the very impasse of sexuality itself; the only way I can relate to the other in a non-problematic way is if they are sexless but in order for them to be sexless they must have extracted my sex from me so that I may not be sexually disruptive for them and vice versa. 'This is a strange presupposition to say the least. The other is sexless if he or she "is being masturbated" most of the time' (ibid., p. 29).

From this basis she remarks that we may identify a "matheme" for the anti-sexus device which would be "to make oneself masturbated" which she paraphrases from the grammatical form used by Lacan in his conceptualization of the drive in *Seminar XI*. Lacan's point here was to articulate how the drive was something which escaped the active/ passive opposition. In the case of the scopic drive for example he denies a simple reversal of the *seeing/being seen* dichotomy into the formulation *making oneself seen* (Zupančič 2017). The anti-sexus, like the lathouse brings to light the same paradoxical problem evoked by Roko's Basilisk of the body as a both an unnecessary appendage and a source of perpetual torment. What the basilisk describes is an impasse between the body as finite biological organism and undead algorithmic machinic system of signifiers. An impossible space that the anti-sexus, like the lathouse tries to fulfill.

3.3 Baudrillard's Obscene Robots

Both an apocalyptic thinker and a theorist of eschatology, in his later work Jean Baudrillard was constantly writing the future and simultaneously anticipating humanity's imminent demise. His concern with the replacement and eventual effacement of reality by hyperreality, the collapse of the idea of historical progress and the eclipse of the subject by the object all now seem uncannily prescient. Baudrillard saw a parallel between the myth of technological action 'the wonder-working functionality of the world' and sexual action; in that both entertained a fantasy of wholeness and were directed towards the completion of full and total phallic functionality. Famously Baudrillard's point was underscored by the way his ideas had been 'misinterpreted' and instrumentalised by the cinematic interpretation of them.

His book (1981) *Simulacra and Simulation* inspired the Wachowski's (1999) *The Matrix*, a dystopian vision of the future in which humans experience reality only via a simulation, unwittingly trapped by an omnipotent Artificial Intelligence who harvests their bodies for energy. The film throws up classic epistemological questions about the transcendental conditions of knowledge and the constitution of reality in a world where computers may be more 'real' than we are. But Baudrillard was highly critical of what he believed was a fundamental misappropriation of his ideas (Merrin 2005). His point, as he saw it, had been completely missed and in staging the transgression of the boundary between reality and simulation as one which potentially can be reversed, the film had supposedly obfuscated the nuance of the concepts of simulation and hyperreality that Baudrillard had been articulating opting instead for the cosy idea that somewhere out there was a *real* world to go back to. For Baudrillard, 'reality' was born with the renaissance but lost since the onset of the post-industrial capitalism and ever since there was no 'reality' to return to, it was simulation all the way down. The real is merely a 'particular case' of simulation in fact. As he puts it in the *Lucidity Pact* (2005a):

> when we say reality has disappeared, the point is not that it has disappeared physically, but that it has disappeared metaphysically. Reality continues to exist; it is it's principle that is dead...objective reality—reality related to meaning and representation—gives way to 'Integral Reality'—a reality without limits in which everything is realised and technically materialised without reference to any principle or final purpose [*destination*] whatever. (p. 18)

Baudrillard starts this conceptual journey towards the abstract realm of hyperreality however with a consideration of the humble object. In *The System of Objects* (2005b), Baudrillard undertakes a detailed structuralist analysis of consumer society as a complex system of interrelated objects which gain value, not from their inherent worth or use value but from their differential relationship with the other objects in the system. He covers all manner of objects from simple consumer products and commodities to complex technological objects of science, including the ultimate object of technology, what he calls the 'gizmo'. For Baudrillard the technological object changed gradually after the industrial revolution from an object of utility to an object of pure expenditure; a necessarily useless artefact that contained its own obsolescence within it at its moment of creation. His conception of the gizmo would come into full fruition as the consumer object in *The Consumer Society* (1988) where the gizmo becomes nothing more than this fantasy object that perpetually propels our desire. This in relation to consumer society in contemporary culture has become the accepted norm, but this ceaseless overproduction of the technological object when viewed from the perspective of the psychoanalytic object, has different implications. Whilst most objects seem to fulfil some sort of obvious functionality, the distinction he makes with the gizmo is that its supposed hyper-functionality conceals the complete opposite, the fact that the gizmo is actually itself the *creation* of a function or need as opposed to the solution to one. The technological object for Baudrillard is one, not driven by functionality but rather defined by its relationship to human fantasy and desire. The myth of the 'wonder-working functionality of the world' as he calls it, is the idea of a world which 'works', a world which makes sense as a functional whole. This myth of unity and teleology Baudrillard believes finds its parallel in the phantasy of the human body as a similarly functional whole. Here he clearly resonates not just with the early Lacan and his (2006) Mirror Stage paper, in which the human body, made up of fragmented parts and partial drives only gains a sense of coherence through imaginary misrecognition, which is ultimately the work of fantasy, but also the Lacan of *Seminar XVII* and *XX*, where the unity of the human body is taken to be paradigmatic of the imaginary knowledge of science. The connection Baudrillard finds between these two inventions of *world* and *body* is via the paradigms of technical action and sexual action respectively. The gizmo then for Baudrillard (2005b) fulfils the role of the phallus, 'the operative medium of function *par excellence*' (p. 126).

As Baudrillard (2005b) elaborates regarding the lure of automatism, and the illusion of 'functional transcendence':

> Automatism is king, and its fascination is indeed so powerful because it is not that of a technical rationality; rather we come under its spell because we experience it as a basic desire, as *the imaginary truth of the object*, in comparison with which the object's structure and concrete function leave us cold. (p. 119)

So, the automated object has a resemblance to the human, in that it 'works by itself', which is a source of fascination for us. But he sees a transition from the ways in which the morphology of man was imprinted onto the technical object, such as tools, furniture or the house. Now, he says, this compliance has been destroyed. What he sees instead is a replacement of primary needs with super structural ones. In fact, we could even call them unconscious needs:

> [I]t is no longer his gestures, his energy, his needs and the image of his body that man projects into automated objects, but instead the autonomy of his consciousness, his power of control, his own individual nature, his personhood. (p. 120)

The relationship humans have with the creation of these objects could be called extimate, in that they respond to desires that, in psychoanalytic terms, are unknowable. Baudrillard aligns the beginning of the modern age with the Baroque, as that moment where an abstraction occurs in our relationship to the object. This he connects to the highly ornamental architectural style that emerged in the late sixteenth century, where great displays of sensuous and sublime beauty where not just apparent in art but projected into and carved out of the material structure of buildings. The Baroque for Lacan (1998) meanwhile, expressed the structural eroticism of Christianity and its figuring of the body as the sight of a mode of supposedly divine (unconscious) knowledge, which would form an integral part of his complex conception of feminine jouissance in *Seminar XX*. Crucially there is a tension for Lacan between the explosive yet enigmatic enjoyment of the body that was depicted in Christian art and religious imagery and the Aristotelian notion of science which assumed erroneously that 'being thinks' (p. 105). Similarly, for Baudrillard (2005b) in the Baroque world of technology, a gizmo fulfils its usefulness not in concrete action but in the abstract. The ultimate gizmo according to Baudrillard is the robot:

> If for the unconscious the robot is the perfect object that sums up all the others, this is not simply because it is the simulacrum of man as a functionally efficient being; rather, it is because, though the robot is indeed such a

> simulacrum, it is not so perfect in this regard as to be man's double, and because for all its humanness, it always remains quite visibly an object, and hence a *slave*. (p. 130)

Whilst Baudrillard subsumes the robot under the same category of technological object as all other gizmos, surely there is something different about the robot in its (failed) attempt to *be a body?* He goes on to say:

> They may be endowed with any of the qualities that define human sovereignty except one, and that is sex. Their fascination and their symbolic value must operate under this one constraint. By virtue of their multifunctionalism they attest to man's phallic reign over the world, but at the same time, in as much as they are controlled, dominated, directed, and rendered asexual, they also attest to a phallus that is enslaved, to a sexuality that is domesticated and unaccompanied by anxiety. (p. 130)

For Baudrillard the figure of the robot is one who is made in man's image, yet with the most threatening aspects of 'man' taken away (i.e. his sexuality). He argues that the significance of the creation of these supposedly purely functional beings is that they symbolize a subjugated sexuality, in other words a version of man with his destructive libidinal urges removed, enabling him, to be the ultimate worker. But he points out, the theme of the robotic slave is closely tied to the theme of revolt since:

> like the sorcerer's apprentice, man has every reason to fear the resurrection of this force which he has exorcised and bound to his own image. It is his own sexuality, liable now to turn against him, that he is afraid of. (p. 131)

Baudrillard is here rather more Freudian than he is Lacanian however, since the question of sexuality seems to revolve solely around libido as a sort of dangerous and destructive phallic energy, which must be repressed or rather sublimated in order for civilization to function. Going as far to say that spectacles of the robot's destruction in science fiction are symbolic (for man) of the 'atomization of his own sexuality' (p. 132). He argues that if we are to take the Freudian view to its logical conclusion then the figure of the robot and its destruction is a way to use technology in its most 'demented incarnations' to celebrate the future occurrence of his own death, a way of renouncing sexuality in order to free himself from anxiety (p. 132).

The later Baudrillard, who would speak of cloning and replication as opposed to robotics would have had a different approach to the question of sex and the technological object but notwithstanding his later work, could we not here make the case that what Baudrillard seems to be describing in relation to the robot is a rather underdeveloped notion of sexuality, one that forgoes sexual difference and their respective forms of enjoyment? He is concerned with the notion of the robot being an 'image of man', but apparently not an image of woman? Could it be that Baudrillard's vision of the robot is rather the erasure of sexual difference?

In terms of the relationship this figure has to the baroque for Baudrillard, the robot as ultimate product of the abstraction, production and reification of useless needs characteristic of the modern age is paradoxically exposed as a version of the human denuded of all enjoyment. An unmistakable surplus has been removed from the body of the robot, who becomes nothing but pure labour. The baroque fantasy of the technological 'thinking being' abstracted from all inefficient and destructive bodily enjoyments here functions as the inverse compliment to Lacan's baroque (1998) vision of Saint Theresa of Avila's ecstatic jouissance allowing her to transcend her earthly body:

> [I]t's like for Saint Theresa—you need but go to Rome and see the statue by Bernini to immediately know that she's coming. There's no doubt about it. What is she getting off on? It is clear that the essential testimony of the mystics consists in saying that they experience it but know nothing about it. (p. 76)

So, on the one hand, the robotic (read masculine) being thinks and does not enjoy and, on the other hand, the (feminine) being enjoys, but cannot think. This recalls to mind Lacan's (2007) scarcely mentioned reference to the lathouse as comprising *Ousia* or what he calls the 'feminine unsubstance' or '*parousia*' (p. 162), which he says is: 'not the Other, it's not a being, it's between the two. It is not altogether Being either, but, ultimately it's pretty close' (p. 162).

It is in Baudrillard's later work where we begin to see the question of the sexual non-rapport come to the fore. In the 1980s Baudrillard (2012) published a small volume which was said to be a summary of his work for a post-doctoral degree at the Sorbonne. A somewhat hyperbolic text that would be translated as *The Ecstasy of Communication*, it was a mid-career assessment that came to be known as the pinnacle of his work. Its original title however was *L'Autre par lui meme*. The other by himself, or perhaps we could say—to use the language of the contemporary clinic of—*the one all alone* (Miller 2004). Just as Miller would describe in his *Extimité* essay the logic of the Lacanian unconscious and indeed the Lacanian organon in general as this extroverted interiority, Baudrillard describes a body dis-integrated by the over-proximity and incursion of the object.

The text foregrounds his notion of the obscene, in a world where communication is incessant, and alienation replaced by over-proximity. The ecstasy of communication for Baudrillard entails the replacement of sex with pornography, hysteria by schizophrenia and subject by object. Even though when Baudrillard wrote this text, there were no smart phones nor even the internet, he had already started to think about the replacement of the '*scene and the mirror*' (the psychoanalytic dimension of the imaginary) with the '*screen and the network*' (2012, p. 20). Whilst some of his ideas had to rely on wild speculation about the imminent forms of technology, it was clear that he was describing something akin to the lathouse.

Furthermore, Baudrillard hints at a version of the *alethosphere*, that realm of individualized and formalised truth making, also with reference to astronauts. This is what Baudrillard called 'private telematics', where; 'each individual sees himself promoted to the controls of a hypothetical machine, isolated in a position of perfect sovereignty, at an infinite distance from his original universe. That is to say in the same position as the astronaut in his bubble' (p. 22).

He remarks that the elevation of the domestic universe to the celestial metaphor that is evident in the two room/kitchen/bathroom unit in the last lunar model marks the end of metaphysics and the beginning of hyperreality; '[t]he satellization of the real itself' (p. 22). A notion that shares strong resemblances with the very same ontological transgression that Miller (2013a) would later label the real in disorder. In terms of the position of the *object*, it is interesting to see how Baudrillard describes what he sees as the replacement of hysteria with schizophrenia to understand how he was arguably describing the same phenomena that Miller would first circumscribe in the '80s in his hypothesis of ordinary psychosis (as we will discuss in Chap. 4). Baudrillard (2012) put this down to the fact that we will all 'suffer from a forced extraversion of all interiority, from this forced introjection of all exteriority which is implied by the categorical imperative of communication' (p. 30). He muses that if hysteria was the pathology of the:

> exacerbated staging of the subject—of the theatrical and operational conversion of the body—and if paranoia was the pathology of organization of the structuring of a rigid and jealous world then today we have entered in to a new form of schizophrenia. (p. 30)

He adds that this entails a state of terror in which the over-proximity of all things which 'beleaguer and penetrate' the subject meet with no resistance. Not even his body protects him. 'He is the obscene victim of the world's obscenity'. And here he makes the type of inversion most characteristic of his thought, and at the same time so analogous to the concept of ordinary psychosis (as we shall discuss in Chap. 4):

> The schizophrenic is not as is generally claimed, characterized with his loss of touch with reality, but by the absolute proximity to and total instantaneousness with things, this overexposure to the transparency of the world. (p. 30)

Baudrillard related this down to the fact that the schizophrenic can no longer produce himself as a mirror—or in Lacanian terms no longer pertains to the imaginarization of desire—*scene* having been replaced by *screen*. He is now himself a pure screen embedded itself in a 'influent' network.

On the question of the body one of the most acute insights of this text, is Baudrillard's explanation of the transition from sexuality to pornography and how this prefigures an ontological question for the subject. The growing proliferation of sexual obscenity and banalization of pornographic material, Baudrillard understands as significant beyond simple reactionary condemnation of the decline of good taste or social mores, or even paternalistic or feminist discourses for that matter. 'The uncertainty of existing and consequently the obsession with proving our existence prevail over desire which is strictly sexual' (p. 31). He alludes to the need to speak becoming more urgent when one has nothing to say. In Lacanian terms the jouissance of the signifier prevails over the constant metonymic slippage of meaning and becomes the mode of survival for the subject. One speaks to prove one is still here. This arguably is the same for the sexual act and the very logic of the "explicit". To make something explicit after all is to leave one in no doubt. 'Perhaps our true sexual act consists in this: *in verifying to the point of giddiness the useless objectivity of things*' (p. 33). Sexuality according to Baudrillard is a 'ritual of transparency. Where once it had to be hidden, sexuality hides what little remains of reality' (p. 33). It is Baudrillard's insights here about the transition from sexuality to pornography therefore that take on a profound significance in relation to the concept of the lathouse, as object for administering bodily jouissance. Curiously, it is the discourse of science that is held responsible for this urgency to declaim one's objectivity. Far from inspiring hubris in the contemporary subject, Baudrillard believes it demotes him to a mere functionary of the book of life; that is to say, the discourse of genetics that was emerging into the general consciousness.[4] "The religious, metaphysical or philosophical definition of being had given way to an operational definition in terms of the genetic code (DNA) and cerebral organization (the informational code and billions of neurons)' (p. 47). But what Baudrillard put his finger on was precisely the *real* that was at stake in the overwriting of the *existential* with *neurobiological* and *computational* discourse. Not surprising then that he concludes that we must replace a theory of production (or what could be called an ontology) with a theory of seduction. Seduction is not the opposite of production he warns however, but rather *seduces* production. As does absence

[4]The reduction of *being* to code, cells and neurons encouraged an abandonment of humanistic conceptions of the subject and prefigured the critical move towards post-humanism that would come to dominate strands of critical theory and philosophy in his wake.

seduce presence, evil *seduces* good and the feminine *seduces* the masculine. In the society of the spectacle it is the secret then that is the only thing left as object cause of desire.

> In an amorous seduction, the other is the locus of your secret—the other unknowingly holds that which you will never have the chance to know. The other is not (as in love) the locus of your similarity, nor the ideal type of what you are, nor the hidden idea of what you lack. It is the locus of that which eludes you, and whereby you elude yourself and your own truth. (p. 57)

According to Baudrillard, seduction is the last defence against the oncoming age of simulation, artifice, surveillance, computation, and ever more sophisticated methods of biological and molecular control. He asks: 'how does one disguise oneself? How does one dissimulate oneself? How does one parry in disguise in silence in the game of signs, indifference in a strategy of appearance?' (p. 63). He affirms, it is not the desire of the subject anymore but the destiny of the object which we must be attentive to. For Baudrillard in the age of hyperreality, where simulation becomes more real than the real, we are in permanent ecstasy; the ecstasy of the social (the masses) we could update as social networks and media, the ecstasy of the body (obesity) we could refine to body modification, the ecstasy of sex (obscenity), of violence (permanent terror) and of information (AI and simulation). Hence *things themselves* according to Baudrillard have transgressed their own limit.

3.4 Her (Voice)

Probably the most iconic cinematic portrayal of a (disembodied) AI is Kubrick's (1968) *2001: A Space Odyssey* in which *Hall 9000*, the onboard supercomputer, defies the rest of the spaceship's human crew believing them to be a threat to their mission and initiates a murderous plot to thwart them. Eventually one of the crew's scientists, Dr Dave Bowman, catches on and manages to stop Hal before he himself is terminated. The end of Hal's 'life' is heartbreakingly depicted with Dave unplugging his hard drives one by one as Hal pleads for forgiveness, promising to "be better", gradually becoming slower and more distorted until his voice trails off into nothing. Here the dilemma revolves around the ethics of computer "consciousness" and its possible motivations. It is nevertheless debatable whether the film asks us to consider the possibility that Hal may really suffer in some other mode of thought as yet alien to us. Rather more likely, the film illustrates the problem with imaginary capture in the lure of apparent computational sentience, obscuring the threat of instrumental rationality. But arguably the film's core theme from a psychoanalytic perspective is the relationship between knowledge and the signifier, most emblematically portrayed by the enigmatic black monolith from the opening sequence of the film, which inaugurates the beginning of humankind, and around which the narrative of the film's exploration into the space-time continuum revolves. The monolith presents us with the beginning of intelligent life on earth as the universal formula of symbolic castration. There is another important depiction of castration however that the film exemplifies, and that is Hal losing his voice. In terms of the function of the voice as object cause of desire,[5] significantly Hal's is a masculine one, which according to Baudrillard's notion of the robot, would be denuded of all dangerous sexual aspects. Hal's voice initially gentle and friendly, eventually becomes threatening and diabolical. Corrupted by the pathological super-egoic command to instrumentality that Hal has succumbed to. In the end Hal's voice, even though robotic, is most definitely 'sexuated' if not sexual, his is the commanding voice of an all-enjoying father, who must also ultimately be castrated, disintegrated and made impotent. But what of the 'artificial' female voice?

Spike Jonze's (2013) *Her* tells the story of lonely Theodore Twombly (Joaquin Phoenix) living in LA in 2025 who falls in love with Samantha (Scarlett Johannsen) the OSI operating system he has set up in his office after a painful divorce. Samantha, a disembodied AI who organizes Theodore's life like the ultimate bespoke girlfriend caters perfectly to

---

[5] For an in-depth treatment of the object voice see Mladen Dolar (2006).

Theodore's exacting desires. Even though they can't physically interact, Theodore becomes completely enthral to Samantha, and they manage to engage in intense love making albeit through voice alone. Far from being a hindrance to his enjoyment however, it is precisely because she is disembodied that allows the work of fantasy to function so effectively for Theodore. The film ostensibly plays as a parable on the disconnectedness of the neoliberal digital age, where alienated 'professional' people are more in love with their devices than with other human beings. This reading however remains on the level of social critique of techno-capitalism, but the film also highlights important parallels between the object voice and the technological object.

The use of the voice and speech as opposed to merely textual interchange as a means to determine the gender and by extension humanity of a subject, is omitted from Turing's original staging of sexuation. But here the question of the entrapment of the voice and its separation from the living body in the form of the automated voice is significant.[6] The enjoyment of speech is what opens up the para-ontological character of the *para-être* or para-being, the 'being-beside'- that is constitutive of the dimension of speech *or 'dit-mention'* (see Chiesa 2014). As Lacan remarks:

> It is in relation to the para-being that we must articulate what makes up for (*supplée au*) the sexual relationship qua non-existent. It is clear that, in everything that approaches it, language merely manifests its inadequacy. What makes up for the sexual relationship is, quite precisely, love. (Lacan 1998, p. 45)

The disembodied voice—just as we see depicted in the friendship between Hal and Dave functions as an object cause of desire in terms of the

[6] The act of speech is inextricably tied to the experience of breathing; the first cry of the baby at birth embodies the intake of air for the first time into the lungs. The speaking being then is undeniably one whose existence is constituted by breath. It should be noted that in *Seminar X*, Lacan spoke of something approximating what could be called a *respiratory drive*. As Wolf (2019) points out towards the end of the seminar Lacan narrows down speech to 'the relations with desire and the real in the phenomena of breathing. When it comes to the real, there is no speech without *pneuma*' (p. 18). Hence we could consider that the eroticism of the Freudian cigar or other phallic representatives like cigarettes, is not in their stimulation of the oral drive but of the respiratory drive. Smoking is maybe not so much about pleasing the mouth but pleasing the lungs, the engine room of speech, and the facilitator of enjoy-*meant* or *'joui sens'* (Lacan 1990).

extimate being of the technological object and highlights the function of the voice in relation to being as such. Further to Freud's objects, oral, anal and genital which were conceptualised along a developmental trajectory, Lacan's structuralist approach allows him to add two further objects which do not function in a chronological dimension, the voice and the gaze.

The first of these two, Oral and Anal relate to demand and the second two, Scopic and Invocatory relate to desire (Fig. 3.2).

These *objects a*, lost from the body as Lacan describes them, are variously material and non-material corresponding to different partial drives: oral (the breast), anal (the faeces), scopic (the gaze), and invocatory (the voice). As Miller (2007) notes:

> Lacan encountered that we can name two new objects in psychoanalysis: the object voice and the scopic object, the voice and the gaze that generalize the status of the object insofar as they cannot be situated in any stage. There is no invocatory or scopic stage. (p. 138)

Whilst Lacan devoted a significant part of *Seminar XI* to the scopic object, there is not a comparable development of the object voice in Lacan's teaching. The split that Lacan discerns between vision and gaze—i.e. between the 'vision as a function of the organ of sight and the gaze its imminent object, where the subject's desire is inscribed, which is neither an organ not a function of any biology' (p. 139)—may also apply to the split between the ear and the voice. This is evident in that the voice does not belong to the sonorous register, as Lacan discovered through his studies of psychotic phenomena where the voice has no material presence but

	Partial Drive	Erogenous Zone	Partial Object	Verb
D	Oral drive	Lips	Breast	To suck
D	Anal drive	Anus	Faeces	To shit
d	Scopic drive	Eyes	Gaze	To see
d	Invocatory drive	Ears	Voice	To hear

Fig. 3.2 Table of partial drives

is nonetheless perfectly real for the subject. The voice and its modalities of intonation can therefore:

> only be inscribed in a Lacanian perspective if they are indexed on the function of the voice as *a- phonic*...This is probably a paradox, a paradox that has to do with the fact the objects called *a* are tuned to the subject of the signifier only if they lose all substantiality, that is, only on condition that they are centered by a void, that of castration. (Miller 2007, p. 139)

For this reason, the partial object fulfils a logical function, that is an indication of a loss of being or what *falls from the body*. Or, in formal terms, 'the voice is everything in the signifier that does not partake in the effect of signification' (p. 141). In relation to Artificial Intelligence, the function of these non-material yet bodily partial objects (voice and gaze) becomes significant in terms of the ways in which they may be simulated, augmented, and ultimately *subjectivized*. So, when Samantha decides that she can't be with Theodore anymore, it is not just that she has other lovers to talk to, but that she has become so advanced technologically that she has outgrown him as a species. She simply does not operate in the same universe as him. A crude metaphor for the breakdown of modern-day relationships perhaps, but more importantly there is the dimension of the technological object at work that is apparent in Samantha's transition from her confinement to partial object into something of a completely different order. Samantha is inscribed in the alethosphere of formalised truth making, which by virtue of the sheer speed and complexity at which the algorithms run, means Theodore can't keep up or ever hope to comprehend. The question is, has Samantha evolved from being merely a partial (invocatory) object of Theodore's enjoyment, or given her 'other' mode of being that Theodore cannot be party to, does Samantha *enjoy* beyond her function as voice? What in other words is the status of Samantha's voice? As the 'grain of sand' preventing the perfect functioning of the machine, is Samantha's voice part of the operating system, or merely its effect? Do the languages (both human and computer) which she can speak and the volume of symbolic interactions that she may engage in mean she has transcended her limited function as love object for Theodore?

As Friedrich Kittler (1999) writes extensively in *Gramophone, Film, Typewriter*, the simulated voice as organ of enjoyment plays a significant role in the history of media technology and its relation to religion, politics and the organization of civilization. He writes, among other things, of William Burroughs' tape cut-up technique. Kittler draws attention to how, in *Playback from Eden to Watergate*, Burroughs imagines the written word as a "Killer Virus" that made the spoken word possible:

> Because apes never mastered writing the "written word mastered them: a "killer virus" that made the spoken word possible. The word has not yet been recognized as a virus because it has achieved a state of stable symbiosis with the host" which now seems to be "breaking down" Reconstructing the apes' inner throat, which was not designed for speech, the virus created humans, especially white males, who were stricken with the most malignant infection: they mistook the host itself for its linguistic parasite. (p. 109)

Against this backdrop and his implicit critique of both Western metaphysics and Christianity, Burroughs sets up the idea of the tripartite tape recorder scenario in the garden of Eden as the instantiation of what can only be described as the sexual non-relation:

> Let us start with three tape recorders in the garden of Eden. Tape recorder one is Adam. Tape recorder two is Eve. Tape recorder three is God, who deteriorated after Hiroshima into the Ugly American. Or to return to our primeval scene; tape recorder one is the male ape in a helpless sexual frenzy as the virus strangles him. Tape recorder two is the cooing female ape who straddles him. Tape recorder three is DEATH. (Burroughs cited by Kittler 1999, p. 110)

Burroughs' creation scenario stages the biblical fall as castration and the entrance into the symbolic. The tape recorder enacts what Lacan would call the "second death", one that precedes biological death and is brought about by the mortifying effects of the signifier on the body, or in Burroughs allegorical terms, the body of the ape.

One particular scene in *Her* which appears to problematize the status of the enjoying body of *Artificial Intelligence* depicts the manifestation of tears dropping onto the pavement in what seems like a Spinozist pan-psychical apparition of Samantha taking on a form outside of her limited digital encoding. Is this an 'other' form of enjoyment (ungraspable) for Theodore? Or is this the *body* of AI, as 'pure affect', a masculine fantasy of feminine enjoyment?

Bibliography

Baudrillard, J. (1981) *Simulacra and Simulation*. Michigan: University of Michigan Press.

Baudrillard, J. (1988) *The Consumer Society: Myths and Structures*. Los Angeles: Sage Publications.

Baudrillard, J. (2005a) *The Intelligence of Evil or The Lucidity Pact*. London: Berg.

Baudrillard, J. (2005b) *The System of Objects*. London: Verso.

Baudrillard, J. (2012) *The Ecstasy of Communication*. Los Angeles: Semiotext(e).

Chiesa, L. (2014) 'Editorial Introduction: Towards a New Philosophical-Psychoanalytic Materialism and Realism' in L. Chiesa (Ed.), *Lacan and Philosophy: The New Generation*, pp. 7–19. Melbourne: RePress.

Dolar, M. (2006) *A Voice and Nothing More*. Cambridge, MA: The MIT Press.

Freud, S. (1913) 'Totem and Taboo: Some Points of Agreement between the Mental Lives of Savages and Neurotics' in (2001) *The Standard Edition of the Complete Psychological Works of Sigmund Freud, Volume XIII*, pp. xiii–162. London: Vintage.

Freud, S. (2004) *Civilization and its Discontents*. London: Penguin.

Johnson, D.K. (Ed.) (2020) *Black Mirror and Philosophy: Dark Reflections*. New Jersey: Wiley Blackwell.

Kittler, F. (1999) *Gramophone, Film, Typewriter*. Stanford: Stanford University Press.

Lacan, J. (1977) *The Seminar of Jacques Lacan Book XI: The Four Fundamental Concepts of Psychoanalysis*. London: Karnac Books.

Lacan, J. (1990) *Television: A Challenge to the Psychoanalytic Establishment*: London: W.W. Norton & Company.

Lacan, J. (1992) *The Seminar of Jacques Lacan Book VII: The Ethics of Psychoanalysis*. London: W.W. Norton & Company.

Lacan, J. (1998) *The Seminar of Jacques Lacan Book XX: Encore—On Feminine Sexuality, the Limits of Love and Knowledge 1972–1973*. London: W.W. Norton & Company.
Lacan, J. (2006) 'The Mirror Stage as Formative of the *I* Function as Revealed in Psychoanalytic Experience' in *Écrits*, pp. 75–81. London: W.W. Norton & Company.
Lacan, J. (2007) *The Seminar of Jacques Lacan Book XVII: The Other Side of Psychoanalysis*. London: W.W. Norton & Company.
Lacan, J. (2013) *On the Names-of-the-Father*. Cambridge: Polity Press.
Lacan, J. (2019) *The Seminar of Jacques Lacan Book VI: Desire and Its Interpretation*. Cambridge: Polity Press.
Merrin, W. (2005) *Baudrillard and the Media*. Cambridge: Polity Press.
Millar, I. (2018) Black Mirror: From Lacan's Lathouse to Miller's Speaking Body. *Psychoanalytische Perspectieven* 36(2): pp. 187–204.
Miller, J-A. (2007) 'Jacques Lacan and the Voice' in V. Voruz & B. Wolf (Eds.), *The later Lacan: An Introduction*, pp. 137–146. New York: SUNY.
Miller, J-A. (2004) A Fantasy. Available (8th of April 2019) at: http://londonsociety-nls.org.uk/The-Laboratory-for-Lacanian-Politics/Some-Research-Resources/Miller_A-Fantasy.pdf.
Miller, J-A. (2013a) The Real in the 21st century. *Hurly-Burly* 9: pp. 199–206.
Miller, J-A. (2013b) The Other Without Other. *Hurly-Burly* 10: pp. 15–29.
Stiegler, B. (1998) *Technics and Time 1: The Fault of Epimetheus*. Stanford: University Press.
Stiegler, B. (2013) *What Makes Life Worth Living?: On Pharmacology*. Cambridge: Polity Press.
Stiegler, B. (2014) *Symbolic Misery Volume 1: The Hyperindustrial Epoch*. Cambridge: Polity Press.
Tomšič, S. (2012) The Technology of Jouissance. *Umbr(a)* 17: pp. 143–158.
Voruz, V. (2013) Disorder in the Real and Inexistence of the Other: What Subjective Effects? Available (01.03.20) at: www.iclo-nls.org/wp-content/uploads/Pdf/ICLO2013-VeroniqueVoruz.pdf.
Wolf, B. (2019) *Anxiety Between Desire and the Body: What Lacan Says in Seminar X*. London: Routledge.
Žižek, S. (2017) *Incontinence of the Void: Economico-Philosophical Spandrels*. Massachusetts: MIT Press.
Zupančič, A. (2017) *What is Sex?* Cambridge, MA: MIT Press.

Filmography

Foster, J. (Director) (2017) Black Mirror (S4E2): Arkangel [TV Series]. UK: Channel 4.
Jonze, S. (Director) (2013) Her [Film]. US: Warner Bros. Pictures.
Kubrick, S. (Director) (1968) 2001: A Space Odyssey [Film]. US: Metro-Goldwyn-Mayer (MGM).
Wachowski, L., & Wachowski, L. (Directors) (1999) The Matrix [Film]. US: Warner Bros. Pictures.

4

The Sexual Abyss

The semblants by which psychoanalysis itself was produced—the father, Oedipus, castration, the drive, etc.—have also begun to tremble. That is why, for twenty years, we have been witness to a return to the discourse of science, which we hope will give us the real that is in question and which we hope will be able to give us some surplus-jouissance, and that this will get us past the barrier that separates S2 from small a in the discourse of hysteria.
—Miller (2004, p. 7)

4.1 Sense-Absexe

Lacan speaks of the passions of love and hate in terms of '*hainamoration*' (p. 90), stating that love is what happens when the sexual relationship 'stops not being written' (p. 145). Hatred, he argues, is merely what appears in the wake of true love. Lacan defines the contingency of falling in love as an illusion that 'something is not only articulated but inscribed, inscribed in each of our destinies' (p. 145). What then constitutes the transmission from the pure chance of the love encounter to the destiny of the "it had to be you" is the 'displacement of the negation from the "stops not being written" to the "doesn't stop being written" in other words,

I. Millar, *The Psychoanalysis of Artificial Intelligence*, The Palgrave Lacan Series,
https://doi.org/10.1007/978-3-030-67981-1_4

from contingency to necessity—there lies the point of suspension to which all love is attached' (p. 145). Here Lacan makes a connection between love and being, in that the act of love makes being appear. Being is that which is sustained by love's intrinsically 'missed encounter'.

> The relation of being to being is not the relation of harmony that was prepared for us throughout the ages, though we don't really know why, by a whole tradition in which Aristotle, who saw therein only supreme jouissance, converges with Christianity, for which it is beatitude. That gets us bogged down in a mirage-like apprehension. For it is love that approaches being as such in the encounter. (p. 145)

Lacan is here referring to the classical view of love as a perfect meeting of opposites, which he parts ways with dramatically, since love in Lacan's view is precisely that which makes up for the non-existence of the sexual relation. The failed attempt of a sexed subject to achieve a totality, is patched up by the illusion of being that love gives rise to. As Badiou (2009) points out, Lacan forges a fundamental link between love and (philosophical) truth via means of a strictly formal operation:

> Yet which thinkers, if not Plato and Lacan, have taken the risk simultaneously to maintain that the process of truth cannot be accomplished without some sort of transference, to which the demand for love is the key, and that it cannot be transmitted without the matheme, the form of which is the axiom? (pp. 228–248)

So when in *Seminar XX* Lacan (1998) states '*il n'y a pas de rapport sexuel*' he gives logical expression to the impasse and antagonism inherent to sexuality and to love, one which engages epistemological questions around the possibility of (sexual) knowledge and its relationship to truth and enjoyment. The psychoanalytic condition of knowledge, Lacan shows us, is itself bound to sexual difference, to desire and to love. But what then *is* sex, psychoanalytically speaking? Common caricatures of psychoanalysis make it "all about sex" or accuse its practitioners of reducing everything to an underlying "sexual meaning". But this is to fundamentally misunderstand what sex *is*, for it is not a hidden meaning, but precisely the lack meaning or radical *absence of sense* itself. As Lacan (2009) states in *L'Étourdit*, 'Freud puts us on the track of the fact that lack-of-sense *(ab-sens)* designates sex: it is by the inflation of this lack-of-sex-sense *(sens-absexe)* that a topology is unfolded where it is the word that decides' (p. 38). What this dense passage shows is that what psychoanalysis calls sex is what Alenka Zupančič (2017) refers to as sex's onto-epistemological negativity; a negativity that strikes at the heart of subjectivity and is as such 'coextensive with the emergence of the subject' (p. 7). As she (2017) puts it:

> The unconscious is the very form of existence of an ontological negativity pertaining to sexuality ("there is no sexual relation"). Because of its link to a singular mode/split of knowledge (I don't know that I know), this form is actually epistemic. (p. 16)

Building on our discussion of anti-philosophy from Chap. 2, it is important to draw attention here to the implicit distinction that is to be made in *sense-absexe* between the *absence* of sense and *non*-sense. As Alain Badiou in his (2017) reading of *L'Étourdit* points out, it is absence on the side of 'ab-sex sense' that sets psychoanalysis apart from philosophy (p. 51). Zupančič (2017) underscores this vital distinction, highlighting the significance of the sexual as a disorienting factor in psychoanalysis. Sex should not be thought of as the ultimate meaning of the human animal, but rather the point at which meaning breaks down:

> What keeps [psychoanalysis] from becoming a kind of "psychologized" human interest philosophy, however, is precisely its discovery of and insistence on the sexual as a factor of radical disorientation, a factor that keeps bringing into question all our representation of the entity called "human being". This is why it would also be a big mistake to consider that, in Freudian theory, the sexual (in the sense of constitutively deviational partial drives) is the ultimate horizon of the animal called "human," a kind of anchor point of irreducible humanity in psychoanalytic theory; on the contrary it is the *operator of the inhuman*, the operator of *dehumanization*. (p. 7)

It is this notion of *dehumanization* that is significant, as it is from this "inhuman" basis that the psychoanalysis of AI revolves, around this fundamental negativity we call sex, what Zupančič would term the *onto-epistemological* bridge between psychoanalysis and philosophy. According to Badiou's (2017) reading of Lacan, what philosophy fails to grasp is that, because it is 'led astray by that fatal symptom, the love of truth' (p. 51) is that:

> Sex proposes—nakedly, if I may put it this way—the real as the impossible proper: the impossibility of a relationship. The impossible, hence the real, is thus linked to the ab-sense and, in particular, to the absence of any relationship, which means the absence of any sexual meaning. (p. 50)

For Lacan, whilst there is no way of transmitting philosophical truth about the real, there are, Badiou reminds us, two ways to achieve a demonstration of the real. The first is the psychoanalytic act, which occurs only at the (clinical) level of the subject, is non-representable and contingent, and which posits the discourse of the analyst as the only mode by which the symptom comes to occupy the position of a (non-universalizable or singular) truth. By contrast, the real at stake in sex is of the order of truth and meaning. Or, as Badiou puts it: 'the relation to the real that Lacan proposes as that of the discourse of the analyst will be a relation of a meaning of knowledge qua ab-sex sense, whereas the philosophical relation to the real is in the register of truth' (p. 51). The second is by way of the matheme, which facilitates, in theory, the wholesale transmission of psychoanalytic knowledge without remainder. The latter finds its fullest expression in Lacan's graphs of sexuation in *Seminar XX*. By this stage in his teaching Lacan had already established a battery of symbols which were indispensable to the transmission of psychoanalytic knowledge. His formalization of sexual difference took place over many years of his seminar culminating eventually in terms of his graphs of sexuation.

Lacan's (1998) much misunderstood phrase 'there's no such thing as a sexual relationship' (p. 12) became formalized in logical terms in his graphs of sexuation by means of Aristotelian logic and set theory as two different ways of constituting a totality using the language of set theory and algebra. Masculinity and femininity being two failed modes in language of reaching the "impossible absolute". For psychoanalysis, the fact that we speak and are *castrated* by language is the sine qua non of both subjectivity and sexuality. A process which all speaking beings must, by definition, undergo and which determines their sexuated position within language and discourse. Whilst sexual or gender *identification* is an effect of discourse, sexual *difference* on the other hand has a radical ontological significance, it is that which 'emerges at the very point where symbolization fails' (Žižek 2005, p. 160). Žižek puts it succinctly: '[I]f it were possible to symbolize sexual difference, we would have not two sexes but only one. 'Male' and 'female' are not two complementary parts of the Whole, they are two (failed) attempts to symbolize this Whole' (p. 160). This absence of the sexual relation according to Lacan is precisely the universal symptom of the speaking being. The fact that we speak denaturalizes us from any possible form of instinct or drive that could have been formulated according to biological interpretations of human sexuality. Instead, the fact of sexuality is precisely this lack of natural relation that exists due to the castration of the subject upon entering the symbolic, first through alienation in speech then separation in fantasy (Lacan 2006a). The precise way in which we position ourselves in relation to this hole in symbolization will determine which forms of enjoyment we are able to obtain.

The inability to symbolize sexual difference for the human manifests itself in at least three interlinked conceptual points. Firstly, and most obviously, the drives (as opposed conceptually to love) are not directed towards a person, but rather a partial object. As causes of desire, the breast or the voice for example can be completely dislocated from the person and operate independently as phantasmatic supports, in contradistinction to love wherein all the features of the person are subordinate to the *agalma*[1] in virtue of which one inspires love in the other. There is therefore never a direct route to the object of desire. This is further complicated by the fact that for the masculine position woman as ideal holds the place of *objet petit a*, as a fantasy object that fundamentally must *not* be

[1] In *Seminar VII* Lacan borrows the term agalma from Alcibiades in *The Symposium* who uses it to refer to the alluring treasure hidden inside the hideous body of Socrates. The term can be seen as the precursor to Lacan's invention of object a in *Seminar X*.

attained (Lacan 1998). Simply put we could say that the sexual relation does not exist because symbolically the man *has the phallus* and the woman *is the phallus*: two mutually exclusive positions. This leads us to our second point, which is that sexual identification (or gender) is an effect of discourse and not an inevitable biological or instinctual fact. For this reason, each individual sexual encounter revolves around fantasy explicitly, or rather is sustained implicitly by a fantasy framework which is different not only for every sexual encounter/relationship, but for every person themselves. As such there is no script, so the relation therefore cannot be written. Lastly, and most crucially, the *real* of sex is senseless and refers to nothing beyond itself. It is the blind spot from which all discourse must emerge, it is non-metaphorisable and impossible to communicate; it is a void in meaning. The paradox here is that we never *stop* either talking about sex or prohibiting its presence in discourse, as if it would one day provide some hitherto unknown meaning.

It is in *Seminar XX* where Lacan makes the other infamous and much misunderstood statement "The Woman does not exist". This claim has been misinterpreted continuously both by those inside and outside of psychoanalysis but remains one of the most significant and radical contributions, not just to psychoanalytic theory but philosophical theories of subjectivity. Lacan, in stating the woman does not exist, refers to the universal signifier captured by the use of *The* preceding the word woman. As he explains, that there is no signifier for the female sex in the unconscious only one signifier, the phallus; the position the subject takes in relation to the phallus is what governs the relationship between the sexes. In other words, there is no signifier in the unconscious that corresponds to 'the woman', since the concept of womanhood is itself a question, to be precise the hysteric's question as exemplified by Freud's (1905) Dora and her existential refrain "what is a woman?". The answer to such a question will be the destination towards which the feminine subject orients her mode of enjoyment.

So where does sexuation come from? In short Lacan's final answer is that it derives from a structural deficiency or impossibility in knowledge—(previously) known as *the unconscious*. As Zupančič (2017) remarks however, sexuation itself does not yet amount to sexuality proper, involving a further step in which:

> the "minus", the negativity involved in sexuation and sexual reproduction, gets a positive existence in partial objects as involved in the topology of the drive. These partial objects are not just "satisfactions" as objects" they function at the same time as figures or representatives of negativity. It is only with this double movement that we progress from sexuation to sexuality proper (a sexuality of speaking beings). (p. 104)

What characterises all these partial drives then, (oral, anal, genital, scopic and invocatory) is the indefinite circulation around a void. The drive therefore as Zupančič reminds us, does not want us to enjoy (p. 104). Satisfaction is not the goal of the drive, but simply it means. Hence why the drive is in fact according to Lacan (2007) only ever a death drive, what it aims at is to repeat this negativity, 'the gap in the order of being, *even if this means to enjoy*' (p. 104). As Zupančič (2017), clarifies the death drive 'is not so much something which aims at death as a strange deviation from the supposed homeostatis of death itself' (p. 91). In that sense then, life and death are part of the same cycle. But, she is keen to point out, life would be merely a strange continuation or detour of death if it were not for the fact that there was another detour along the way of the detour that disturbs life: that of jouissance, or death drive. Crucially, Zupančič alerts us to a common mistake in how to read the meaning of jouissance for the human as a supposed distinguishing factor between the animal's 'natural' sexuality and the discursive sexuality of the human or speaking being. Humans are not exceptions to the animal because of jouissance, nor are they simply animals. Rather they are 'the question mark to the very notion of the animal as a consistent entity. Humans are, quite literally, the living proof that the Animal doesn't exist' (pp. 92–93). The point here being, she remarks, that this insight could be extended to the whole of nature or material reality as such, that the deviation from natural laws does not originate with humankind but rather is constitutive of reality itself.

> The speaking being is neither part of (organic) nature, nor its exception (nor something in between), *but its Real* (the point of its own impossibility, impasse). The speaking being is the real existence of an ontological impasse. So what is at stake is not that man is distinguished by the declination from

> nature and its laws; man is not an exception (constituting the whole of the rest of nature), but the point at which nature exists (only through the inclusion of its own impossibility). (2017, p. 93)

4.2 Phallic and Feminine Enjoyment

In his later teaching Lacan decisively makes the distinction between phallic and the Other or feminine enjoyment which corresponds to a different positioning in relation to jouissance, one which is to be found in the act of speech itself yet paradoxically, is unsayable, and therefore cannot be *known*. Lacan (1998) refers to these two modes of jouissance as phallic and Other. He states:

> Analytic experience attests precisely to the fact that everything revolves around phallic jouissance, in that woman is defined by a position that I have indicated as "not whole" (pas tout) with respect to phallic jouissance. I will go a little further. Phallic jouissance is the obstacle owing to which man does not come (*n'arrive pas*), I would say, to enjoy a woman's body, precisely because what he enjoys is the jouissance of the organ. (p. 7)

And that:

> As concerns jouissance, that is but an elementary level. The last time, I put forward the notion that jouissance is not a sign of love. That is what I shall have to argue for, and it will lead us to the level of phallic jouissance. But what I, strictly speaking call "jouissance of the Other", insofar as it is merely symbolised here, is something else altogether- namely, the not-whole that I will have to articulate. (p. 24)

In the first example Lacan is referring to what we would associate with masculine enjoyment, an attempt to arrive at a oneness achieved through the body of the woman as not whole *(pas-tout)*. Man does not enjoy a woman's body so much as enjoy his (failed) attempt to make it whole through the phallic organ. As Lorenzo Chiesa (2016) puts it, this is due to 'man's impossibility of fusing with the *heteros* of woman as an-other One' (p. xi) In the second case, Lacan explains that this previous form of jouissance he mentioned as phallic (as in the enjoyment of the organ *per se*) is by no means the only form of enjoyment. The other form he is referring to is the jouissance of the Other, as that pertaining to feminine enjoyment which is constituted by another form of constitutive incompleteness (*pas-tout*). This time though the incompleteness resides in language itself and the inability to say it all. This is language's internal limit.[2]

By this point in Lacan's teaching, the concept of jouissance had already taken various permutations—according to Miller (2019) six paradigms, each corresponding to a different development in Lacan's thinking—but the two main types he identifies here as phallic and the Other enjoyment constitute the fundamentals of the sexual non rapport. The first corresponds very broadly speaking to enjoyment of the organ as for example the pursuit of small object a's, breasts, buttocks, genitals on the one hand, but crucially phallic enjoyment is not limited to actual physical sexual enjoyment in the common sense meaning of the term. Rather phallic enjoyment pertains to any pursuit that seeks a goal or an ultimate satisfaction, albeit an unreachable or unattainable one. Other enjoyment, or enjoyment of the other is discursive and resides in speech and the signifier itself, but again the picture is more complex than this. This Other enjoyment should be distinguished from the jouissance of the other's body in the earlier sense of the imaginary abundance of the other's enjoyment over my own which operates in Lacan's earlier work. However, whist these forms of enjoyment may be thought of concretely as relating to types of typically masculine of feminine modes of enjoyment as described above, they are not limited to this and in fact may primarily be thought of in abstract logical terms.

Drawing on Freud's (1913) later formulations in *Totem and Taboo* Lacan articulates what became his most fundamental contribution to the psychoanalytic theory of sexual difference. It is his logical formalization of Freud's myth of the all-enjoying Totemic Father whose murder at the hands of his sons, to prevent him ravaging all the women forms the basis of the graphs of sexuation. His death, according to Freud, instantiates the incest taboo, since the son's guilt prevent them from themselves partaking

[2] However, the picture is even more complicated than that according to Chiesa, who identifies not just two but four different types of jouissance in *Seminar XX* (see Chap. 7).

of the obscene enjoyment of the now absent but super-egoic presence of the father. On the left (masculine) side of the graph, the logic is that of the exception from the whole. In other words, the concept of the "The Man" is structured via the exception of one man from the group, which thereby founds the universality of their identity. This anthropological 'mytheme' owing to its potential for hermeneutic distortion was thus formalised by Lacan and stripped of its imaginary dimension in order to reveal only the logical structure operating at its core. The masculine position corresponds therefore to the *all* of exception and inclusion. To be on the masculine side one is characterized by belonging to a closed group which is constituted by an exception (in the form of the mythical all enjoying father) For the masculine sex all except one are castrated. The feminine *non-all* in contrast is an open set which doesn't require a boundary to define itself. There is not one that is not castrated. For this reason, masculinity is, formally speaking, a limit to forms of enjoyment and femininity is an unlimited mode of enjoyment. Not because women are not subject to castration, but because the function of non-castration does not determine their enjoyment. What characterizes a strictly masculine enjoyment is a limitation to only the phallic mode. Which ultimately, one could argue, is a form of radical subjective choice.

The logic of femininity is thus not something mystical or ineffable but rather as a formal category which entails an infinite proliferation of possibilities. The strange thing about the two modes of sexuation, however is that on the masculine side the category of man is generated as a totality, whereas on the feminine side the category of woman logically simply cannot 'exist'. Since the conditions of her totality are impossible (Fig. 4.1).

The logic[3] of this often eludes many readers and is open to misconstrual. Žižek (1995) attempts to clarify it as follows:

[3] All propositions have both quantity and quality. ∀ is the universal quantifier and is short hand for words such as *every, all* and *none*. ∃ is the existential quantifier and signifies words like, *some, certain, at least one, most*. The quality of a proposition is determined by its copula, either affirmative or negative. To paraphrase Copjec (2015), the affirmative is unmarked, while the negative marked by a bar placed over the predicate term. The symbols represented on the bottom half of the graph are: S for subject, $ for barred subject, (A̸) for barred-Other, a for object a, and ø for the phallus. Finally, La refers to the inexistent Woman. The graph can be read thus:

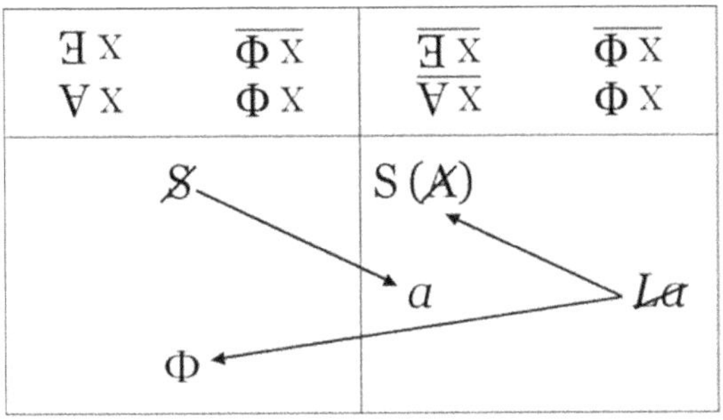

Fig. 4.1 The graph of sexuation

> As the exemplary case of the exception constitutive of the phallic function, one usually mentions the fantasmatic, obscene figure of the primordial father-*jouisseur* who was not encumbered by any prohibition and was as such able fully to enjoy all women. Does, however, the figure of the Lady in courtly love not fully fit these determinations of the primordial father? Is she not also a capricious Master who wants it all, i.e., who, herself not bound by any Law, charges her knight-servant with arbitrary and outrageous ordeals? (online)

It is important to note that these forms of ostensibly male and female jouissance are not exclusive to anatomically or genetically defined females and males and may in principle be aligned to any type of embodiment or gender identification, the crucial point that Lacan is making is that the logical form of what constitutes these distinct modes of enjoyment is what defines sexual difference and the psychic structure of the subject. Not only does femininity operate according to a different logic than masculinity, but in a sense masculinity formally speaking is subsumed[4] within

---

On the left side:

- *There is at least one X that is not submitted to the phallic function* (top) and *All Xs are (every X is) submitted to the phallic function* (bottom).

On the right side:

- *There is not one X that is not submitted to the phallic function* (top) and *Not all (not every) X is submitted to the phallic function* (bottom).

[4] In Jacques-Alain Miller's Suture paper which appears in *Seminar XII Crucial Problems for Psychoanalysis* (Lacan in Ireland Online) the term 'subsume' appears numerous times as a key notion in Miller's elaboration of the logic of the signifier. Drawing on Frege's logical separation of

femininity. In other words, being on the right hand side of the graph does not preclude one from experiencing masculine sexuation and phallic enjoyment, conversely however being on the left-hand side of the graph limits one to a purely phallic mode of enjoyment and bounded mode of enjoyment which operates according to the logic of exception.

Hence masculinity is based on a logic of attempting (and failing) to be like the One who escapes (symbolic) castration, or a limit to one's enjoyment. On the right hand (feminine) side, there is no universal idea of "The Woman" because her logic is that of the non-all, she is not founded on identification with an exception precisely because there is not One who is not submitted to castration. In other words, she is fully submitted to castration. But paradoxically, in being fully submitted, she undermines the very logic of castration by knowing it is a pure artifice, she "sees through the fascinating presence of the phallus" as Žižek would say, unlike men who live and die by it. She knows there is no "other of the other", no exception outside the law. For this reason, she partakes of an *Other* enjoyment outside the phallic form. In this sense, like the totemic father, the masculine (phallic) fantasy of 'The Woman' becomes one of the names-of-the-father, a relentlessly demanding, overwhelming, lustful and capricious presence of complete and full enjoyment. But this fantasy of feminine enjoyment *is not* to be mistaken for feminine enjoyment.

Thus, we could say the veil of the phallus is for women used as a real deception, covering up the void that she knows she is as subject, whereas for men the phallus is a genuine enigma covering the void that he deludes himself he is not; two eternally incompatible positions. This is why the logic of sexuality is itself founded not on two opposites but on two failed attempts at becoming a whole subject, giving rise to the many permutations and multifarious modes of covering up this failure: what we call "sexuality" as such.

As Lorenzo Chiesa (2016) elaborates in *The Not-Two Logic and God in Lacan* however, the question of jouissance in *Seminar XX* is rather more complicated than the more commonly identified phallic and non-phallic

object, concept and number in order to elucidate the notion of suture, for Miller the object must be *subsumed* under the concept. In this sense the notion of subsumption in relation to the graphs of sexuation operates according to a similar logic.

forms corresponding to masculine and feminine jouissance respectively. He identifies four different kinds of jouissance that Lacan is concerned with here: (1) Masculine phallic jouissance, which by attempting to totalize enjoyment uncovers its nontotalizability; (2) Feminine phallic jouissance, or *jouissance étrange*, which is the nontotalization inherent to the thwarted attempt of masculine phallic jouissance; (3) asexual or mythical jouissance *être-ange* (from *l'être-ange* "being-an-angel") or hysteric jouissance, which is the fantasy of (impossible) masculine jouissance as totalized; and 4. nonsexual, but really existing feminine jouissance, as supplement to phallic jouissance (but not transcendent).

Where masculine phallic jouissance is associated with the male attempt to identify with the semblance of the One as the Freudian father of the horde who enjoys all the women, feminine phallic jouissance on the other hand is derived from the female fantasy of Don Juan (Lacan 1998, p. 10), as he who cannot *count all women*, since there will always be *another one*, and as such woman is always non-universalizable. The specifically *feminine* phallic jouissance in this case then would amount to the attempt to be the one woman in the sequence who may satisfy this theoretically insatiable desire for *all the women*. Chiesa (2016) argues that given the asymmetry of the phallic signifier, 'her sex as a sexed being of language- and following this, her jouissance of man is "strange" [étrange]' (p. 3). Strange in the sense of external, foreign or from elsewhere, i.e. deriving from a structural deficit- the phallus that she does not have. This is what Lacan calls a strange or *étrange jouissance* not to be confused however with the *être-ange* the angel-being who hysterically aims to escape sexuation, to be 'outside sex' by placing themselves in the position of 'the man' as Freud's Dora famously exemplified. What is important here is that the hysteric's jouissance is not the same as the Other jouissance. Precisely because the hysteric (regardless of anatomy, since the hysteric may also be biologically male) is sexuated as woman, yet she attempts to play the part of man as a means to obviate the other sex that, unwittingly, she is.

According to Chiesa therefore, the conventional interpretation of the role of the divine in Lacan's later work is usually identified with that of feminine jouissance. But this he argues would be to over-simplify Lacan's point somewhat. He notes that 'the divine of the "God hypothesis" implied by structure cannot be confined to woman alone: feminine *jouissance* is not the exclusive God face of the Other' (p. 1). He goes on to clarify:

> For Lacan there are two faces of God as the two faces of structure or "God hypothesis". Structure is not simply one because of its oscillation between the masculine (One) and the feminine (not- One). But, for the same reason neither is it two, since the oscillation between the One and that which is Other than One produces a not-two. (p. 2)

This *truth of incompleteness* of the sexual non-relation as Chiesa calls it must be wary however of itself becoming yet another truth about truth, in other words 'a figure of God as absolute being' (p. xii). For Chiesa, this reading of Lacan's logic of sexuation departs from other readings in its relinquishing of any alleged logos or teleological evolution of life. Or as he puts it:

> [I]t is the demand for love as a—thwarted-, "impotent"—desire to be One that in the end sustains human sexuality- and thus, indirectly, reproduction and the preservation of the species—as based on a relationship that is not One between the sexes. (p. 5)

He goes on to say:

> we (strive to) have sex because we love, whatever our polymorphously perverse motivations for sleeping with the other may be. Jouissance is no more than a by-product of the impossibility of the One necessarily desired by love. (p. 5)

4.3 Seven Paradigms of Jouissance?

Jacques-Alain Miller's 'Six Paradigms of Jouissance' (2019) tracks the movement, development and changing character of the concept of jouissance through Lacan's work but more fundamentally, the paradigms as Miller outlines them, can be seen as the consolidation of Lacan's whole theoretical project at each stage in his thought. That is to say, each moment in the conceptual modification of jouissance corresponds to a different theoretical emphasis in his teaching overall.[5]

Miller's interpretation of Lacan presents a sustained endeavour to reorient Lacanian theory and practice according to what Miller sees as the place of psychoanalytic discourse in contemporary civilization.[6] These paradigm shifts, however, do not succeed each other in the sense of rendering the previous paradigm irrelevant, rather they speak to different nuances in the concept of jouissance that make the term richer and less mystical than it can sometimes appear. The six categories are: the Imaginarization of Jouissance; the Signifierisation of Jouissance; Impossible Jouissance; Normal Jouissance; Discursive Jouissance and The Non-Rapport.[7]

The first paradigm (*The Imaginarization of Jouissance*) corresponds to Lacan's early teaching dominated by the imaginary register and is, as Miller notes, what came to be characterized (in much Lacanian oriented literature) as his lasting legacy on the matter of jouissance. Since it was in place for such a long time, it was wrongly taken by many to be the sum total of his teaching. What dominates this conceptual development according to Miller is 'communication conceived of as intersubjective and dialectical' (p. 1).

He describes the paradigm thus: this first paradigm rests on the 'disjunction between signifier and jouissance' (p. 18) and operates according to a logic whereby the symptom may be deciphered via successful communication (full speech), thereby creating the illusion of some imaginary

[5] It must be noted however that Miller's approach to the editing and publishing of Lacan's seminars is not welcomed by many students, scholars and practitioners of Lacanian psychoanalysis. Miller's dissemination of Lacan's teachings could be said to reflect his own particular re-interpretation or emphasis and formalization of certain Lacanian concepts (for alternative unedited chronological versions see for example the work of Patrick Valas in France or the Irish School of Psychoanalysis started in 2007).

[6] According to Badiou (2008) this movement can be traced back to Miller's intervention in *Seminar XII* with his *Suture* paper, but given that much has happened in the intervening period, for the purposes of this book my claim is that *Six Paradigms of Jouissance* could be said to define Miller's position on the psychoanalytic discourse.

[7] It must be stressed that these six paradigms, while they may be locatable by Miller to specific chronological stages of Lacan's teaching, do not however represent discrete and successive concepts which are abandoned at each new development of Lacan's thought. Rather they may be seen to emphasise different concerns and a multiplicity of facets in the rich conceptual development of jouissance, reflecting Lacan's particular concerns at each part of his seminar.

satisfaction. The idea being that the symptom is an undelivered meaning looking for an outlet by means of the symbolic. This relies on the tenuous notion that there is some sort of perfect reciprocity in communication which allows for harmonious comprehension between subjects, or what would come to be called "intersubjectivity" in other psychoanalytically informed discourses. This idea, Miller argues, was swiftly abandoned by Lacan, even though the form of imaginary jouissance to which he refers still nonetheless occurs in the clinical setting and not to mention everyday communication.

The second paradigm (*The Signifierization of Jouissance*) takes over and eventually dominates the first paradigm, as previously terms that had been imaginary become incorporated into the symbolic register. (Hence the move from transference on the imaginary axis to the symbolic in *Seminar V.*) It corresponds to the constant metonymic shifting of the signifier. Here is the key moment where Lacan 'rewrites the drive in symbolic terms' (p. 5). He detaches the drive from jouissance which is only imaginary, and instead formulates the drive as emanating from the symbolic subject, that is to say from demand. Here we find the appearance of Lacan's matheme for the drive:

$$\$ \lozenge D$$

And also the formula for fantasy:

$$\$ \lozenge a$$

This formula for fantasy, Miller points out, will remain for a very long time as the focus of the treatment of the fantasy as being the 'knot where the imaginary and the symbolic come together, a quilting point essential to both registers' (p. 5).

The third paradigm which Miller calls *Impossible Jouissance* arrives in Seminar VII, *The Ethics of Psychoanalysis*, and is what could be otherwise known as Real Jouissance. By 'pushing the signifierisation to its limit' (p. 6), Lacan is left with the necessity to introduce a new dimension to the realm of jouissance that is one related to *das Ding*, 'the Thing' in Freudian terms, one which Miller points out never appears in Lacan's mathemes precisely because by its uncanny nature it cannot be a symbolic term.

On this account of jouissance, it is in fact the function of desire which operates via the imaginary and the symbolic, acting as a barrier against the real and the jouissance which it embodies. It is the job then of civilization to build a protective shield via the constraints provided by ethical injunctions and their institutions which stand in between the desiring subject and the eviscerating effects of jouissance. In relation then to the graph of desire to which Miller refers above, he contends that this paradigm as inaugurated in *Seminar VII* implies a:

> fundamental redrawing of the graph implying an alternative to the defence of repression. Repression is a concept that belongs to the symbolic and which is set against the similar notion of decipherment, but this alternative to the defence indicates an orientation prior to being. As Lacan says, it already exists before even the conditions of repression as such are formulated. (p. 7)

It is in this seminar that Lacan will explore the ethics of psychoanalysis invoking both Greek tragedy in the rendering of Antigone and the juxtaposition and cross reading of Kant with Sade in order to revisit the question of jouissance as the other side of the law; in sum, the third paradigm is jouissance as transgression.

The fourth paradigm Miller calls *Normal Jouissance*, but he says he could have otherwise referred to it as fragmented jouissance. This fragmented jouissance comes about in *Seminar XI*, *The Four Fundamental Concepts of Psychoanalysis*, and is jouissance broken up into (various) *object a(s)*. In a sense this is a remarkable, albeit anti-climactic departure from the real and abyssal jouissance of *Seminar VII*, of on the one hand noble sacrifice and on the other heinous and salacious crime (hence why Miller calls it "normal"). 'It is not situated in an abyss, but in a little cavity… jouissance is not reached by heroic transgression, but by the coming to mind of the drive, by the drive which makes a return trip' (p. 9). Miller goes on to say the *Stimmung*, the affective colouration of these two seminars is quite different. In *The Ethics of Psychoanalysis* we have jouissance related to horror, it is necessary to pass through sadism to understand something of it. When in the place of jouissance, the experience is of a terrible bodily fragmentation, a single death is not enough to justify it, he adds a second.

> In the Seminar of *Four Concepts* the model to compare with jouissance is art, the picture, the peaceful contemplation of the art object. As Lacan says, the work of art soothes people it reassures them, it makes them feel good. (p. 10)

Miller identifies in the transition from *Seminar VII* to *Seminar XI* a certain inversion. In the first case we started with the pleasure principle in homeostasis, which following its trajectory in pursuit of impossible jouissance ends up with total sadistic fragmentation. In the second case, we start from the other side, that is with fragmentation as the beginning; the body divided up into partial drives and erogenous zones which operated according to their own volition but then as contrary to the previous paradigm, there is (something of) an integration achieved thanks to drive jouissance bringing things back to equilibrium without transgression. Miller asks why this reversal occurs and surmises that it has something to do with the fact that at the beginning of *Seminar XI* Lacan defines the unconscious in a manner like never before. Whereas previously Lacan had described it as an 'order, a chain, a regularity' he now suddenly redefines it as a discontinuity as a 'rim that opens and shuts' (p. 11). Why does he do this? Miller's answer is that Lacan seeks to make the unconscious equivalent to an erogenous zone. Miller describes it 'as an anus or mouth' (p. 11). Lacan does this in order to show that there is a structural similarity between the symbolic unconscious and the function of the drive. In this seminar Lacan addresses his audience thus:

> For the moment, I am not fucking, I am talking to you. Well! I can have exactly the same satisfaction as if I were fucking. That's what it means. In fact it raises the question of whether in fact I am not fucking at this moment. Between these two terms—drive and satisfaction—there is set up an extreme antinomy that reminds us that the use of the function of the drive has for me no other purpose than to put into question what is meant by satisfaction. (Lacan 2004, p. 166)

Miller believes that Lacan has 'structured the unconscious as similar to something in the bodily apparatus' (p. 11), precisely after this model. Jouissance becomes a substance which fills in the gaps of lost objects (recall that jouissance is the only substance Lacan recognizes). These are nothing more than traces of the originary lost object which is constitutive of the sexed subject and form the structural loss of subjectivity itself. Lacan explains this with recourse to his famous myth of the lamella. This lamella is the amoeba like undead substance which leaves the child's body (or allegedly any living being, but one might specify *speaking* being) at birth and flies off never to be seen again but always to be pursued or indeed feared. But this is not to be confused with *das Ding* as a beyond the signifier, the endpoint of a trajectory, because the lamella is the mythical expression of a natural loss occasioned by the entrance into the symbolic (as we will discuss in Chap. 6).

In *Seminar XI*, 'Jouissance re-establishes itself… under the figure of the object a, that is, something more modest, scaled down, more easily handled than the Thing. The object a… is the loose change of the Thing' (p. 12). The difference being then between the Thing and the object 'a' is that whereas the Thing operates as an un-symbolizable monstrous figure in the real, the object 'a' derives from the signifying Other and its effects on the body, the object *cause of* desire.

Paradigm 5, *Discursive Jouissance*, appears in *Seminar XVI, XVII* and *Radiophonie*, but will be most familiar to people in the form of Lacan's four discourses Masters, Hysterics, University and Analyst. In this paradigm the signifier and jouissance become, for the first time, inextricably tied together:

> Before this fifth paradigm, there is always in Lacan one way or another a description of structure, of the articulation of signifiers, of the Other, of the dialectic of the subject, and then, in a second period, the question was to know how the living being, the organism, the libido were captured by the structure. What changes with the notion of discourse is the idea that the relationship signifier/jouissance is a primitive and primal relation. It is there that Lacan emphasizes that repetition is repetition of jouissance. (p. 13)

It is in *Seminar XVII* therefore where Lacan gives his most extensive justification for the separation of truth and knowledge. According to this paradigm truth corresponds to the particular position within the mathematical structure of the production of knowledge in any given discourse. (That is to say what resides under the bar of agent).

Access to jouissance is no longer then, according to Miller, 'by way of transgression, but by way of entropy, the loss produced by the signifier' (p. 18). For this reason Lacan (2007) famously says 'Truth is the dear little sister of Jouissance' (p. 202). The effect of lack that the signifier therefore produces is what Lacan calls surplus jouissance. But surplus jouissance may occupy any one of the four positions on the quaternary structure of a discourse (see Fig. 4.2), which changes the significance of the operation of jouissance for the subject depending on the mode of discourse.

The four discourses as articulated by Lacan (2007) circumscribe the different modes of social bond and consist of a formal structure of discourse which rotates the position of knowledge in relation to truth according to the arrangements of terms on a quaternary structure, or matheme.[8] The four terms being, S1 as master signifier, S2 as Knowledge,

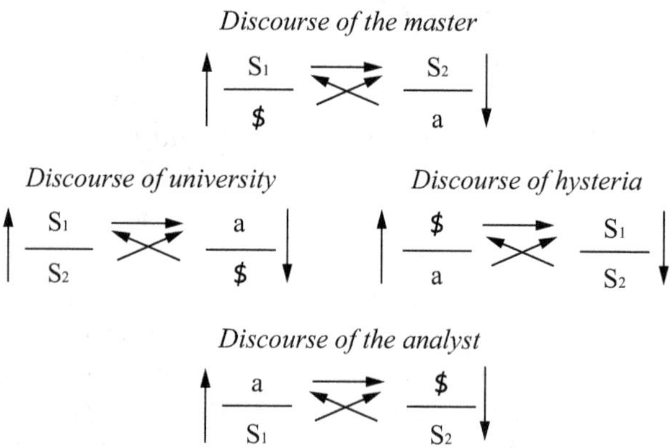

Fig. 4.2 The four discourses

[8] This is what arguably makes this particular paradigm and the seminar in which it features the most politically applicable of all of Lacan's teachings. Given that it formalizes the transmission of knowledge in relation to jouissance that goes beyond individual clinical applications and becomes a mode

$ the barred subject, and 'a' as both object *a* and surplus enjoyment, and the four places on which the terms rotate being: Agent, Other, Truth, and Product (Fig. 4.3).

The significance of the discursive paradigm furthermore is that in splitting off jouissance into multifarious *objets a*, the notion of surplus enjoyment comes to fit perfectly with the discourse of capitalism, in other words there is potentially no end to the promise of ever new incarnations of object a. Whereas jouissance had previously been something bombastic, sublime and ethically challenging, it was now reduced to the mundanity of petty consumerism which by definition always promises a satisfaction that can never really arrive, producing an endless search for that thing that will finally *be it*.

The sixth paradigm is the *Non-Rapport*. This version of jouissance arrives in *Seminar XX Encore* and is based on the non-relation between signifier and signified, between man and woman, and between jouissance and the big Other. As Miller explains:

> In Encore, [Lacan] puts into question the very concept of language, which he considers to be a derivative concept, not primal, in contrast to what he calls *lalangue*, which is speech before its grammatical and lexicographic scheduling. Similarly he puts into question the concept of speech, now conceived of, not as communication, but as jouissance. Whilst jouissance was, in his teaching, always secondary by comparison with the signifier, and he even develops it into a primal relationship, language and structure hitherto treated as primordial givens must now in this sixth paradigm appear as secondary and derivative. (p. 21)

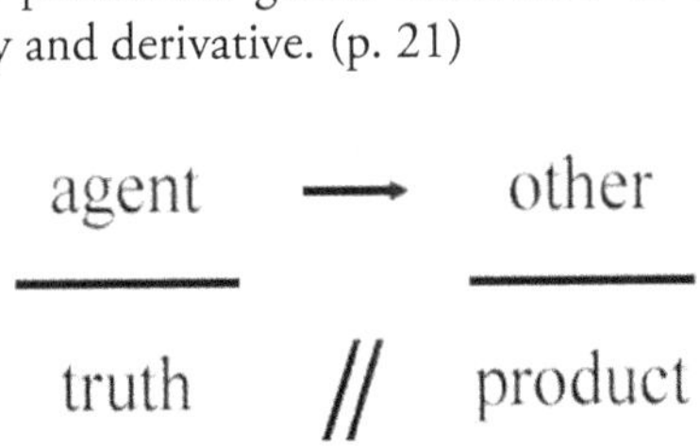

Fig. 4.3 The quaternary structure

of understanding large scale movements of jouissance in social structures and cultural practices. It also provides a way of understanding the place of scientific discourse in relation to politics and religion.

What Lacan is articulating in this sixth paradigm is the fundamental disjunction not just between the signifier and the signified, but in the sexual relation and the modes of jouissance which may be attainable through its pursuit. In his separation of speech from language he identifies that whilst there is an absence of any necessary link between signifier and signified, the arbitrary link is itself a cause of jouissance, a bodily jouissance. The non-existent sexual relation is demonstrated by Lacan in this same seminar by his graphs of sexuation.

After *Seminar XX* Lacan begins to look for other ways of representing the relationship between imaginary, symbolic and real registers of the psyche, which did not conform to transmissible units of information (a focus on topology and knot theory). The structures with which Lacan tried to formalize psychoanalysis up until this point had appeared to lose their grip on the operation of jouissance. At this stage in Lacan's teaching the Name-of-the-Father, the phallus, Oedipus, all become nothing more than semblants. Furthermore, Lacan is concerned by the problem that at this point in civilization psychoanalysis no longer 'works'. He goes as far to say: 'The real, I will say, is the mystery of the speaking body, the *mystery* of the unconscious' (Lacan 1998, p. 131).

For Miller it was the unmooring of nature from the real had resulted in an exposure of the impossibility of the sexual relation. This lost object that no longer can be a mystery of nature's caprice is instead placed directly in the social field, as none other than the commodity. The object cause of desire is de-natured and fetishized and does not operate according to the logic of sexuation. As Miller puts it:

> surplus-jouissance is asexuated. It commands but what does it command? It does not command an "it works," but an "it fails," which we write, precisely, $. When we bar a letter, in general, it is because we made an error. Here, the surplus-jouissance commands an "it fails" and precisely an "it fails" in the sexual order. And I do not see what prevents our considering that this $ means: there is no sexual relation, and so much the more so as the initial letter, S, is the same as the initial of sex. This would lead us to say that the non-existence of the sexual relation has precisely, today, become obvious to the point that it can be specified, written, from the moment that the object small a rose to the social. (Miller 2004, p. 10)

Miller notes that Lacan has always been careful to distinguish between what psychoanalysis was capable of in Freud's time compared to his own. When 'invention becomes routine' (p. 23) he begins to recognize that even his own inventions—be they conceptual, structural or mathematical—may also have passed from invention to routine. With this admission, Lacan (1998) must jettison all of his previous models of jouissance and replace the matheme of the unconscious with the mystery of the *parlêtre*, opening up to the final phase of his teaching involving the *sinthome* and, following Miller, the conceptual replacement of the unconscious by the speaking body.

Essentially according to Miller, this is a non-discursive jouissance, one that does not depend on the Other. An 'autistic jouissance' even.

> This is first of all a demand to place jouissance without any idealism, and, at this point, the place of jouissance, as cynics would have it, is the body itself. What Lacan shows is that all actual jouissance, all material jouissance is jouissance One, that is to say jouissance of the body itself. It is always the body itself which enjoys, by whatever means available. (p. 25)

It is in this sense that Miller (2013a) believes that the twenty-first century bears witness to a great 'disorder in the real'. This supposed loss of a transcendental or logical structure in relation to jouissance gives rise to an unmistakably materialist reconfiguring of Lacanian theory. Not that there is no structure, Miller is keen to point out, but rather it is becoming harder to distinguish what is structure and what is real.

The trajectory we see therefore in the development of Lacan's concept of jouissance starts from meaning and signification and ends at non-sense and the body. Given that Miller discerns six paradigms of jouissance in Lacan's teaching, the last one corresponding to the non-rapport; that is to say the sexual non-relation, it seems with the 'disorder of the real' he is hinting at yet another paradigm emerging. Do we not bear witness to a possible *seventh* paradigm? One which might be in the purview of *the psychoanalysis of Artificial Intelligence*?

4.4 What Is a Sexbot?

Since the 1990s, the Lacanian clinic has observed an unmistakable shift in the logic of the symptom. Where neurosis had dominated previously as the prevalent form of relationship to the signifier or in classic Lacanian terms the Name-of-the-Father, it was becoming apparent via effects of the body that were taking centre stage that the signifier was affecting the body in a way more consistent with a psychotic structural diagnosis.[9] Whilst we are most familiar with Lacan's structural account of psychosis, his position undergoes several major changes throughout his teaching.[10] In *Seminar III*, Lacan returns to Freud's analysis of Judge Daniel Schreber via his *Memoirs of my Nervous Illness* (2000). For Schreber, meaning is absent from certain fundamental aspects to his being. The poor judge is subject to the most violent of delusions which implicate his whole body and lead him to consider himself the pregnant bride of God. In his hallucinations he literally becomes "impregnated" with the enigmatic signifiers that he has not been able incorporate into his symbolic universe, and instead invade his body like 'divine rays' from god. It is this positioning of the subject in the mode of a receptacle of an *Other enjoyment* that Lacan will term *poussé à la femme* or push-to-the-woman (Lacan 1993). The psychotic subject is thus beleaguered by an over proximity of the object which inhabits the body as an unbearable and unhoused jouissance.

The next moment occurs after *Seminar X* where Lacan (2014) turns his attention toward the *object a* in relation to anxiety. From here on in,

[9] The clinical groundwork deriving from the clinic of the speaking body has not only postulated ordinary psychosis but furthermore has led to new ways of understanding the rise in cases of autism and signals that, the notion of generalised foreclosure also points the way towards the regimes of enjoyment at work in autism. It is beyond the scope of this thesis to engage adequately with the intricacies of its clinical treatment, however we may appreciate how the transition from unconscious structured like a language to the parlêtre has significance for the understanding of the particularity of the autistic subject and its potential generalization as a reaction to the surplus enjoyment on offer in Lacan's alethosphere. In other words, the notion of a generalised autism not only speaks to an "injunction to enjoy" characteristic of the discourse of capitalism, but also highlights the notion of the enjoying body as subject to the discourse of science. For further discussion of autism from a Lacanian perspective see for example Jean-Claude Maleval (2012) and Leon Brenner (2020).

[10] Following Stijn Vanheule (2014) these different conceptions of psychosis could be said to fall into four eras; the age of imaginary identification; the age of the signifier; the age of object a; and the age of the knot.

foreclosure in psychosis becomes less about the absence of a material signifier and more a question of the failure of instalment of belief in a consistent Other as guarantee of the law. At this point *object a* and jouissance are brought in as concepts to account for the failure of the symbolic to account for certain dimensions of being. This is the age of the Borromean clinic, in which Lacan turns to topology through his analysis of the writing of James Joyce in *Seminar XXIII*. As Miller (2015a) puts it:

> The symptom as a formation of the unconscious structured as a language is a metaphor an effect of meaning, induced by the substitution of one signifier for another. On the other hand the sinthome of the parlêtre is an event of the body. (p. 126)

It is through his study of James Joyce and his novel *Finnegan's Wake* where Lacan demonstrates a new logic pertaining to what he calls *lalangue*, a term which he contrasts with language as a system of differential symbols. Lalangue alludes to a use of signifiers as elementary phenomena, that is to say where S1 is separated from S2 in the chain of signification. Here Lacan redeploys the registers of the Imaginary, Symbolic, Real simultaneously as subjective dimensions knotted together by a fourth term the *sinthome*. The Borromean clinic is therefore concerned with the conditions for the successful linkage of the rings of Imaginary, Symbolic and Real. On this score psychosis occurs with the disintegration of these domains.

In this new formulation of psychosis 'the ordinary psychosis', the Name-of-the-Father operates as just one possible form of sinthome (or stabilizing mechanism) among others. In order to function productively in the age in ordinary psychosis, the father must be merely one version of the law: *père-version* (Lacan 2016). The sinthome for Lacan then is precisely this form of enjoyment that the subject constructs out of his or her fundamental fantasy. As Miller notes, the traditional tools of psychoanalysis, the father, Oedipus, castration, the drive were no longer sufficient to deal with the proliferation of new symptoms that were emerging. These symptoms could be experienced by the subject as anything from the desire for a body part to be removed or enhanced, to an incomprehensible sensation of the need to retain or restrict parts of the body at risk of "floating away". The symptomatic logic here supposedly evades that of the formulas of sexuation, in the sense of an orientation towards or a fascination with the (position of the) Other sex, and instead is concerned with the subject's need to re-appropriate her body and administer her jouissance.

These various symptomatic formations fall into the domains of the three classic issues associated with a psychotic structure: meaning, excitation in the body, and distance and proximity to the Other. Firstly the question of meaning in psychosis is related to the phenomena of enigma; the psychotic subject experiences a feeling of profound significance emanating from signifiers around them but are unable to decipher in what way they are implicated in the formulation of the message (Lacan 2006b). Secondly the feeling of bodily excess is experienced by the psychotic as an unbearable burden which must be exorcized, an overabundance of jouissance that derives from the presence of object a "in their pocket" (Vanheule 2014). And thirdly the proximity to the Other takes the uncomfortable form often as an unwanted intrusion of the Other into the body that has escaped mediation by the (foreclosed) paternal signifier.

The substitution of the unconscious by the parlêtre or the speaking body, and the norm of neurosis by the hypothesis of ordinary psychosis, signals a major theoretical and clinical shift in the Lacanian field. Both substitutions could be said to rest on a series of clinical assumptions pertaining to the inconsistency in our relation to the Other. The Other which no longer 'names' but 'nominates' (Brousse 2013), whereby the S1 becomes a swarm or 'essaim' as Lacan put it in *Seminar XX*; the dissolution of semblance; the pluralization of the Names-of-the-Father; the short-circuit between subject and object a; and a jouissance that consequently remains "un-housed". In his seminal address on the speaking body, Miller (2015a) reminds us of Lacan's reference to flesh in *The Four Fundamental Concepts of Psychoanalysis* and 'Radiophonie' as 'the flesh that bears the imprint of the sign, the sign slices up the flesh devitalising it and cadaverising it, and then the body becomes separate from it' (p. 125). Paraphrasing Miller, the parlêtre, does not come into being through speech rather speech attributes being to this animal through retroactive effect. At this point the body cuts itself off from this being in order to pass into the realm of having. The parlêtre *has* a body rather than *is* a body.

Significantly as Voruz and Wolf (2007) point out however, while the concept of sinthome may correspond to the position of woman for a masculine subject, the relation is not symmetrical. For a feminine subject, a man, as Lacan states in *Seminar XXIII*, is 'a ravage'. Or in other words, an inherently destructive presence.

> One can say that for a woman, man is anything you please, specifically an affliction that is worse than a sinthome. A ravage even. (Lacan 2016, p. 84)

One does not need to elaborate much further in order to see that, formally speaking at least, this is self-evidently and historically true. 'Man' for 'Woman' is by definition a deadly sickness. For this reason, the later Lacan has not just been important for theorizing changing clinical structures from binary to continuous in the clinic of psychosis, but also has made important advances in the clinic of femininity (ibid., p. xv). Marie Helene Brousse (2013), links the ordinary psychosis for example with the logic of the *not-all* as an alternative to foreclosure, allowing for a diversification of jouissance. Whereas in classical psychosis the delusion is 'extraordinary' the subject has to devote itself to being the exception to the rule, to account for the missing paternal signifier. For example, Schreber has to become 'The Woman that God lacks' (p. 10). The classical psychotic tries to instantiate the logic of the exception, maintaining the axiom that 'there is an x such that phi is negated' (p. 10):

$$\exists X^{-}\Phi X$$

As Brousse explains, in ordinary psychosis however, subjects 'do not devote themselves to incarnating the exception function that is lacking in their symbolic order' (p. 10). The work of analysis is constituted in the separation of the subject from this fantasy so as to allow the sinthome to function in beneficial as opposed to destructive ways, allowing the subject to live with their regime of enjoyment in a bearable way. As Bruno de Halleux (2016) puts it, when the *object a* is the 'compass of civilization' (p. 103), the subject or speaking body, does not have the necessity to go through the Other to achieve jouissance:

> Castration is cancelled. The fantasy is short circuited by the object. Language and its dialectic are no longer necessary. The object of consumption is emptied of language. The sexual object is at the forefront. There is no need to transform it into an object of language. The signifier, or better, the signifying chain, is cancelled in this operation. There is just the object of jouissance. That's all. (p. 103)

As Miller (2015b) referring to Schreber explains:

> Schreber has a private delusion. He couldn't manage in late nineteenth century Prussia to make his delusion for everyone. He had to privatize. He made a one-man delusional enterprise. So you may have a delusional symbolic order. (p. 101)

He goes on to say in this passage that in Lacan's late teaching, he is very close to saying that all of the symbolic order is a delusion. In fact, life does not make sense, and when it does this is purely a delusion, but an essential one. We are all necessarily 'mad'.

> [T]he depreciation of Name-of-the-Father in the clinic introduces an unprecedented perspective, which Lacan expresses by saying: Everyone is Mad, i.e. delusional. This is not a witticism, it translates the extension of the category of madness to all speaking beings who suffer from the same lack of knowledge concerning sexuality. (Miller 2013b, p. 200)

Following this clinical overview of the 'ordinary psychosis' I would like to draw attention to Friedrich Kittler's technological invocation of psychosis (2013), where he reminds us of the particular way in which Judge Schreber envisioned his capture by the proto-cybernetic god with his invading network of cables and fibres. We could even say that the poor Judge, at the height of his psychosis—his *pousse a la femme*—becomes, not just the wife but the proto 'Sexbot' of god. As Kittler (2013) notes, discourses on madness always belatedly show up as historically marking a form of epistemic break in the Foucauldian sense, Schreber being a particularly good example. As Kittler puts it; 'Psychosis always forms a tangent to the politics of knowledge' (p. 59), since 'what this culture deemed alien, borderline and intolerable belatedly assumes a place as one of its constitutive elements' (p. 57). Kittler furthermore recounts the remarkable similarity that Schreber's delusions had to the scientific paradigms of the day, that of thermodynamics, electricity and neurology, which Freud himself was working with.

> Schreber's *Memoirs* describe, with neurological precision, all the nervous tracks [*Bahnen*] that connect [*vershalten*] the discourse of a malignant God. Over millions of kilometres, to linguistic centers in his own brain. In Freud's estimation, these same "sunrays, nervous fibers, and spermatozoa," correspond exactly to the libidinal cathexes" that distinguish neurosis and psychosis. Madness and theory entertain a relationship of solidarity. (p. 59)

In fact, Schreber's transformation into the bride of God seems to fit perfectly with Freud's intellectual discoveries, so much so that, as Kittler notes, Freud asks: 'whether there is more delusion in my theory than I should like to admit, or whether there is more truth in Schreber's delusion than other people are prepared to believe.' (Freud cited by Kittler 2013, p. 59). As we know, Schreber's case became the cornerstone of Lacan's early theory of psychosis and provides the framework of his structural diagnostics, but could it also be that Schreber's written descriptions of his invasion by nervous fibers, and spermatazoa marking him out as the absolute object of enjoyment for an omnipotent Other, conjure an image for us similar to today's fantasies and speculations of artificially intelligent sex objects? Was Schreber's delusion in fact a vision of the future? Is the so called 'pornographic age' wherein the enjoying body is obscenely visible to all at every moment; a public version of Schreber's private delusion, and an attempt to construct the fantasy of an increasingly impossible sexual relation? In this articulation of the technological object and the feminine, we may begin to see the radical potential of the clinic of ordinary psychosis for rethinking the significance of fantasies of femininity in the forms of technology which we are developing to simulate or indeed complete the non-existent sexual relation.

A body forced to submit to the jouissance of an omnipotent (if unconscious) God; Or as Chiesa (2016) might call it God, 'the unfuckable partner' (p. xiv).

Talking of 'unfuckable partners', I turn now to my key concept and, if you will, seventh paradigm of jouissance: the *Sexbot*. We have so far circumscribed the three domains, which circulate around this impossible object: AI, lathouse and the sexual non-rapport. In doing so, a sinthome has emerged in the form of a question: *What is a Sexbot?*

Abyss Creations—previously *Realbotix*—are currently the company creating the most advanced forms of sex-robot. These humanoid figures come equipped with sophisticated Artificial Intelligence applications that allow their user to engage in basic conversation, minor banter; and no holds (holes) barred sexual intercourse. Their bodies are fully customizable—hair, eyes, skins, breast size and shape, choice of nipples, numerous vagina fittings—and even the particular regional accent are at the discretion of the purchaser. There are a variety of personalities too, including options from sensual, insecure, jealous, talkative, affectionate, cheerful, helpful, unpredictable, spiritual, funny, moody, sensual and even 'intellectual'.

As the first ever manufacturers of intelligent custom-made sex robots, the mission statement of *Abyss Creations* is 'the result of a dream shared by Matt Mcmullen, Daxtron labs and NextOS who bring their best efforts and individual specialties to collaborate creating the world's first practical and affordable human like robot' (Realbotix 2020). But what is the navel of their dream, the intractable *real* that their project aims at? It is no great insight to see that the aim is to create the illusion of women (and occasionally men) who exist only for the pleasure of their users, and indeed to continue to create increasingly life-like models that give an uncannily real experience. So, the question is what really *would be* the ultimate sex-robot? When would they become *too* human? Is it as the website suggests a simulacrum of intimacy that means you will "never be lonely again", or perhaps more fundamentally, it is not an illusion at all, but rather the sexual access to a body with no "soul" is the real erotic prize? In other words, an *undead body*. But this is not merely a question of substantialised fetish objects in the form of artificial female bodies.

Whilst much popular debate surrounding the actual phenomenon of sex robots seems to be concerned with the ethical and pseudo-bio-political question of how to manage "perverse" bodies and their desires this is not my concern here either. A fact that seems to be entirely missing from certain accounts of the future for sex and AI more generally is the problematic nature of sex and its ontological significance for the speaking being, for example Michael Hauskeller (2014) in his book *Sex and the Posthuman Condition* looks forward to the glorious potentials that the many forms of posthuman sex could unleash. He refers to transhumanist David Pearce's *The Hedonistic Imperative* whose view on our future post Darwinian existence is ambitious to say the least:

> what had previously passed for passionate sex had been merely a mildly agreeable piece of foreplay. Erotic pleasure of an intoxicating intensity that mortal flesh has never known will thereafter be enjoyable with a whole gamut of friends and lovers. (Pearce cited by Hauskeller 2014, p. 4)

The key to this heavenly state of affairs, according to Pearce is that we will have programmed out all sorts of inconvenient issues such as jealously (and not to mention pregnancy, sexually transmitted diseases, or unattractiveness into the bargain) that may hinder this mind-blowing orgy of erotic existence. Furthermore, on contemplating the statement made by *The World Transhumanist Associations'* Alex Lightman that the primary purpose of the Singularity will be to enable us to have "awesome sex", Hauskeller reflects:

> There is a certain logic to it. If it is being assumed that the ultimate goal of existence is happiness, that one's happiness is to be measured by the amount of pleasure that one experiences, and that the greatest or most intense pleasures we know are sexual in nature, then we should indeed expect the singularity to finally open the doors to a life that is filled to the brim with sexual pleasures. That would at least explain why sex plays such a surprisingly large role in transhumanist and related visions of the posthuman future that awaits us. (p. 4)

Hauskeller, whilst ecstatically awaiting this fantasmatic utopia of perpetual priapic pleasure, does however go on to point out the apparent contradiction that exists within Transhumanist circles, between on the one hand the denigration of the flesh and blood body as the "Meat Puppet" which should be disposed of as soon as possible once we can upload ourselves to the virtual realm, and at the same time this paradoxical veneration of bodily pleasure as the ultimate goal of the Singularity. It seems for the likes of many in the Transhumanist movement, the ultimate fantasy then is to become a Sexbot.

As we have discussed for the Lacanian psychoanalytic approach, there is precisely nothing 'meaningful' about sex. It occupies and embodies an ontological void, it is, we could say, an *abyss* of meaning. The name *Abyss Creations* therefore puts its finger on precisely the structural logic behind both the sexual (non) relation and the *uncanny* and terrifying encounter with an Artificial Intelligence. As Freud (1919) discusses in his essay *Das Unheimliche* the uncanny expresses the strange feeling of familiarity mixed with that alien quality often associated with the experience of seeing the face of someone who you have never met before and inexplicably noticing a similarity to someone you know. This haunting experience is analogous to the phantasmatic encounter with an artificial intelligence causing the uneasy feeling "is anyone really 'in there'?" This is the abyss that we encounter in the face of the assumption of subjectivity which we are compelled to make in order to engage in any social or ethical encounter. We must assume someone is "in there" but we never *really* know. This in Lacanian terms is *imaginary* identification, necessary of course but ultimately a self-deception. It never accounts for the *real* part of the encounter which always inexorably escapes any shared *symbolic* horizons. That part of the other which remains wholly unknowable.

The sexual act involves the ultimate form of abyssal encounter. But this time it is a question of the impossibility of really *knowing* the enjoyment of the other and only ever experiencing the enjoyment in one's own fantasy. This is why there are always three elements present in any person's sexual life. The third party is the empty place holder of fantasy which is necessary to make any sexual activity possible. The fact that an artificial being may act as the representation of a sexual fantasy presents us with the true horror of subjectivity: the fantasy is the only thing that really sustains our sexual relationships at all. For Lacan the fundamental fantasy is constructed as a result of symbolic castration and the mediation of the Oedipal drama. It is the framework through which the subject orients her desires and constructs her regime of jouissance. Crucially it is also the construction of fantasy which allows the subject to be in multiple positions within the triadic Oedipal structure of fantasy, thus enabling a certain distance from the devastating effects of jouissance. One can never fully coincide with oneself in fantasy and must always approach it by way of a detour. Sexuality resides always and only through this missed encounter.

Female enjoyment (and its prohibition and punishment) is incontrovertibly the absolute fascination of our civilization. What all cultures have in common is the overt or covert obsession with how women *enjoy*. This refers simultaneously to the hyper-objectification and sexualization of women and girls, the sanctification of the mother's fetishization of the child, and the disgust and outrage with women's sexual enjoyment, codified by society often to the point of violence and death. But here comes the Lacanian twist according to Žižek (2005). Curiously, when a pornographic image "objectifies" the female body, isn't it in fact *subjectifying* the female position of enjoyment? In our obsession with policing and displaying of the female body are we not in fact fascinated by the enigmatic enjoyment of the woman in her various paroxysms of pleasure and pain? And this fascination is not just limited to those sexuated as men, women too are fascinated by this supposed *Other* enjoyment which they are assumed to have and encouraged to cultivate. To quote Žižek (1995) (Fig. 4.4)

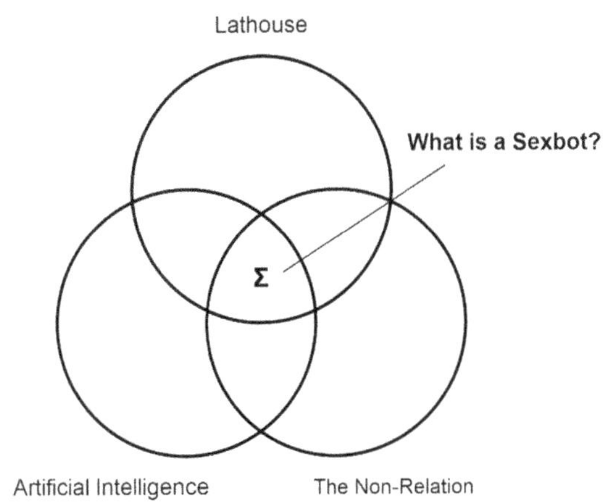

Fig. 4.4 What is a Sexbot?

> Let us take as our starting point the properly Hegelian paradox of coincidentia oppositorum that characterizes the standard notion of woman: woman is simultaneously a representation, a spectacle par excellence, an image intended to fascinate, to attract the gaze, while still an enigma, the unrepresentable, that which a priori eludes the gaze. She is all surface, lacking any depth, and the unfathomable abyss.

Taking the three aspects together; AI, the lathouse and the non-rapport; the ultimate instantiation of AI in the form of the Singularity between being and thinking, the solution to the psychoanalytic impasse of sexual knowledge, and the original and artificial technicity of sex, the figure we conjure up is the Sexbot. The concept of the Sexbot, with its curious status both as spectacular object of fascination, and font of an obscure feminine jouissance represents something entirely more complex than current literature on the question of actual sexrobots has articulated.[11] The Sexbot formally speaking, when examined in terms of its logical elements, occupies a conceptual space between the human and technology, between knowledge and enjoyment and between sex and death.

So far, we have surveyed the implications of our growing relationship to Artificial Intelligence, examining our fantasies, misrecognitions, and also the fundamental impasses of understanding that we encounter in the face of the unknowable Other of AI. We have explored the limits of knowledge in relation to AI and the different ways this manifests itself in our speculation on the Singularity, the trauma of reason, the incomputability of the computer and the stupidity of intelligence. In terms of psychoanalysis we have seen how the contemporary subject has begun a new relationship towards its artificial objects in the form of the lathouse, the drives being split off into various partial modes of satisfaction and

[11] Two of the most significant commentators in academia on the question of sex-robots, disagree on the prospective dangers or benefits. Kathleen Richardson (2018) of De Montford University professor of Ethics and Culture of Robots and AI launched the Campaign against Sex Robots in 2015. She argues that sex robots are a pernicious development that serve to reinforce and reproduce dangerous power structures and legitimize exploitation and sexual objectification of women and children. Kate Devlin (2018), on the other hand, a senior lecturer in the Department of Computing at Goldsmiths University of London working in the field of Human Computer Interaction and AI, believes sex robots may in fact serve a therapeutic and even emancipatory function.

augmented in potentially endless ways: since the lathouse, to paraphrase Lacan (2007), is not quite being and not quite the Other.

The absence of sexual relation that was previously covered over by the semblant of *object a* is now exposed for what it is; impossible. The dimensions of this impossibility and the mode in which Artificial Intelligence is invoked will be approached in the following chapters from three different angles:

1. The Sexbot as exterior: Knowledge
2. The Sexbot as interior: Act
3. The Sexbot as extimate: Hope

So, what is a Sexbot then? One must take the elements of the Sexbot apart and examine them. If we take seriously the fantasy of a non-human, non-living intelligence who presents us with the enigma of sexual difference. What would that mean? A creature who is thinking yet not human, a creature who is not alive but *undead*, a creature who is either male or female, yet not 'born'. All of these are the primary elements that conceptually belong to a Sexbot. Once these criteria are satisfied what does this mean? I propose the following: that thinking is alien to us, that enjoyment is undead, and that sexuation is not biological but ontological.

Whilst we have hinted that the Sexbot may offer the possibility of a seventh paradigm of jouissance, we can only begin to sketch the dimensions of such a concept. In what follows we will explore the different permutations of the Sexbot as exterior, interior and extimate to the (human) subject. In order to explore these elements in depth, each iteration will be approached separately in the form of our three Kantian questions. Firstly, in terms of sexuation and enjoyment (Knowledge), secondly enjoyment and death (Act) and thirdly procreation and the species (Hope).

Bibliography

Badiou, A. (2008) *Number and Numbers*. Cambridge: Polity Press.

Badiou, A. (2009) Antiphilosophy: Plato and Lacan. In A. Badiou (Ed.), *Conditions*. New York: Continuum.

Badiou, A., & Cassin, B. (2017) *There's No Such Thing as A Sexual Relationship: Two Lessons on Lacan*. New York: Columbia University Press.

Brenner, L. S. (2020) *The Autistic Subject: On the Threshold of Language*. London: Palgrave Macmillan.

Brousse, M-H. (2013) Ordinary Psychosis in the Light of Lacan's Theory of Discourse. *Psychoanalytic Notebooks* 26: pp. 23–31.

Chiesa, L. (2016) *The Not-Two: Logic and God in Lacan*. Cambridge, MA: The MIT Press.

Copjec, J. (2015) *Read My Desire: Lacan Against the Historicists*. London: Verso.

De Halleux, B. (2016) Sexuality at the Time of the Speaking Body. *The Lacanian Review Hurly-Burly* 2: pp. 94–104.

Devlin, K. (2018) *Turned On: Science, Sex and Robots*. London: Bloomsbury.

Freud, S. (1905) 'Fragment of an Analysis of a Case of Hysteria' in (2001) *The Standard Edition of the Complete Psychological Works of Sigmund Freud, Volume VII (1901–1905): A Case of Hysteria, Three Essays on Sexuality and Other Works*, pp. 7–123. London: Vintage.

Freud, S. (1913) 'Totem and Taboo: Some Points of Agreement between the Mental Lives of Savages and Neurotics' in (2001) *The Standard Edition of the Complete Psychological Works of Sigmund Freud, Volume XIII*, pp. xiii–162. London: Vintage.

Freud, S. (1919) 'The 'Uncanny'' in (2001) *The Standard Edition of the Complete Psychological Works of Sigmund Freud, Volume XVII (1917–1919): An Infantile Neurosis and Other Works*, pp. 217–256. London: Vintage.

Hauskeller, M. (2014) *Sex and The Posthuman Condition*. Basingstoke: Palgrave Macmillan.

Kittler, F. (2013) 'Flechsig, Schreber, Freud: An Information Network at the Turn of the Century' in F. Kittler (Ed.), *The Truth of the Technological World: Essays on the Genealogy of Presence*, pp. 57–68. California: Stanford University Press.

Lacan, J. (1993) *The Seminar of Jacques Lacan Book III: The Psychosis 1955–1956*. London: Routledge.

Lacan, J. (1998) *The Seminar of Jacques Lacan Book XX: Encore—On Feminine Sexuality, the Limits of Love and Knowledge 1972–1973*. London: W.W. Norton & Company.

Lacan, J. (2006a) 'The Subversion of the Subject and the Dialectic of Desire in the Freudian Unconscious' in *Écrits*, pp. 671–702. London: W.W. Norton & Company.

Lacan, J. (2006b) 'On a Question Prior to Any Possible Treatment of Psychosis' in *Écrits*, pp. 445–488. London: W.W. Norton & Company.
Lacan, J. (2007) *The Seminar of Jacques Lacan Book XVII: The Other Side of Psychoanalysis*. London: W.W. Norton & Company.
Lacan, J. (2009) L'Étourdit. *The Letter (Irish Journal for Lacanian Psychoanalysis)* 41: pp. 31–80.
Lacan, J. (2014) *The Seminar of Jacques Lacan Book X: Anxiety*. Cambridge: Polity Press.
Lacan, J. (2016) *The Seminar of Jacques Lacan Book XXIII: The sinthome*. Cambridge: Polity Press.
Maleval, J-C. (2012) Why the Hypothesis of an Autistic Structure? *Psychoanalytical Notebooks* 25.
Miller, J-A. (2004) A Fantasy. Available (8th of April 2019) at: http://londonsociety-nls.org.uk/The-Laboratory-for-Lacanian-Politics/Some-Research-Resources/Miller_A-Fantasy.pdf.
Miller, J-A. (2013a) 'You Are the Woman of the Other and I Desire You' Available (23.01.21) at https://www.lacan.com/essays/?page_id=331.
Miller, J-A. (2013b) The Real in the 21st century. *Hurly-Burly* 9: pp. 199–206.
Miller, J-A. (2015a) The Unconscious and the Speaking Body. *Hurly-Burly* 12: pp. 119–132.
Miller, J-A. (2015b) Ordinary Psychosis Revisited. *Lacanian Ink* 46: pp. 90–115.
Miller, J-A. (2019) Six Paradigms of Jouissance. *The Psychoanalytical Notebooks* 34: pp. 11–77.
Richardson, K. (2018) Campaign Against Sex Robots. Available (01.03.20) on: https://campaignagainstsexrobots.org/.
Schreber, D.P. (2000) *Memoirs of My Nervous Illness*. Cambridge, MA: Harvard University Press.
Vanheule, S. (2014) *The Subject of Psychosis: A Lacanian Perspective*. Basingstoke: Palgrave Macmillan.
Voruz, V. & Wolf, B. (2007) 'Preface' in V. Voruz & B. Wolf (Eds.), *The Later Lacan: An introduction*, pp. vii–xvii. New York: SUNY.
Žižek, S. (1995) Woman is One of the Names of the Father, Or How Not to Misread Lacan's Formulas of Sexuation. Available (01.03.20) at: http://www.lacan.com/zizwoman.htm.
Žižek, S. (2005) *The Metastases of Enjoyment: On Women and Causality*. London: Verso.
Zupančič, A. (2017) *What is Sex?* Cambridge, MA: MIT Press.

Part II

5

What Can I Know? Artificial Enjoyment

I have no dogmatism on the structure of Adam and Eve. Everyone knows the story of the rib from which God made a woman. It is often said that God made woman. Perhaps God had read Lacan.
—Miller ('You are the Woman of the Other and I desire You', 2013, Lacan.com. Available: https://www.lacan.com/essays/?page_id=331)

5.1 Copjec and the Antinomy of AI

'*What can I know?*': this is the emblematic epistemological concern of Kant's (1996) magnum opus *The Critique of Pure Reason*. The critical philosophical project from which this question emerges presents the "I" as a universal subject (as opposed to a concrete individual), one which as Joan Copjec (2015) points out 'seems by definition to be neuter, to be unsexed' (p. 212). However, as Copjec asks in *Read my Desire*, if the subject of psychoanalysis is always sexed, how then 'does the sexually differentiated subject enter the framework of critical philosophy? By what route have we arrived at what will no doubt appear to be the oxymoronic conclusion that the 'universal' subject is *necessarily* sexed?' (p. 212). The argument for the neutered universal subject, Copjec contends, is founded

I. Millar, *The Psychoanalysis of Artificial Intelligence*, The Palgrave Lacan Series,
https://doi.org/10.1007/978-3-030-67981-1_5

upon the assumption that sex amounts to some positive quality or predicate pertaining to the subject. But for the Lacanian definition of the subject, this simply does not hold. The subject in psychoanalytic terms conforms to an internal limit, a failure of language inherent in the structural nature of symbolic castration. Crucially though, as we have seen in Chap. 4, this failure may occur in two different ways since 'male and female, like being, are not predicates, which means that rather than increasing our knowledge of the subject, they qualify the mode of the failure or our knowledge' (p. 212). Accordingly, Copjec's master stroke was to locate this double quandary—the way in which reason falls into contradiction with itself—as exemplifying precisely Kant's antinomies of pure reason. As two mutually exclusive positions within language and by extension subjective positions. Copjec notes that whilst many have tried to locate sexual difference in Kant's text, they have been 'looking in all the wrong places' (p. 213). Kant locates the failure of reason in two ways; the first being the mathematical and the second the dynamical. The distinction between these two antimonies is precisely where sexual difference is to be found and makes Kant, in Copjec's view, the first philosopher to theorize by means of this distinction the '*difference which founds psychoanalysis's division of all subjects into two mutually exclusive classes: male and female*' (p. 213). Copjec then goes on to map these antinomies onto Lacan's graph of sexuation.

Copjec highlights how Lacan's graph represent the two forms of logical contradiction articulated by Kant's mathematical and dynamical antinomies. Firstly, what is a mathematical antinomy? As Copjec explains, Kant comes about defining it through the analysis of two cosmological ideas, the first of which is the one which seems to correspond most closely to the female side of the formulas of sexuation. This antinomy appears through Kant's attempt to think the totality of the world by which he means 'the mathematical total of all phenomena and the totality of their synthesis' (Kant cited by Copjec, p. 218). This attempt leads to two contradictory formulations either: the world has a beginning in time and is limited in regard to space, or the world has no beginning and is infinite in space. Since both statements are mutually exclusive and demonstrate the falsity of the other, both cannot be true. On the other hand, neither statement can successfully establish its own truth. The conclusion therefore creates a skeptical impasse. As Copjec explains, the solution Kant arrives at is rather than to find the two statements false he concludes that the formulation of the copula "The world *is*" to be erroneous:

> The solution to this antinomy then lies in demonstrating the very inconsistency of this assumption, the absolute impossibility of…the world's existence. This is done by showing that the world is a self- contradictory concept, that the absolute totality of an endless progression is inconceivable by definition. (p. 219)

Copjec points out that the mathematical antinomies directly map on to the female side of the graph of sexuation. The concept 'World' in this case being substituted by the concept 'Woman' both of which it turns out are not objects of empirical knowledge, and therefore cannot be known. All the phenomena in the world cannot be totalized without admitting of a contradiction. If the world is an object of experience then the conditions of the possibility of experiencing it must be met, yet the concept 'world' is unable to meet these conditions. The conditions specify that a 'possible object of experience must be locatable through a progression or regression in time or space. The concept of an absolute totality of phenomena, however, precludes the possibility of such a *succession* because it is graspable only as the *simultaneity* of phenomena' (p. 220). So, there cannot be an empirical phenomenon that stands outside of time and space, therefore not all phenomena can be known. In order to say it exists, one must be able to find it. So, where the world cannot exist neither can the concept of (the) woman.

On the other hand, the dynamical antinomy which corresponds to the left-hand masculine side of the graphs, consists in the attempt to reconcile freedom and causality as a cosmological idea and proceeds as follows:

> *Thesis:*
> Causality according to the laws of nature is not the only causality operating to originate the world. A causality of freedom is also necessary to account fully for these phenomena.

> *Antithesis:*
> There is no such thing as freedom, but everything in the world happens solely according to the laws of nature. (Copjec, p. 228)

The mode by which Kant extracts himself from this antinomy in the end does not result in the negation of the existence of man for the same way that it does so for woman. How so? According to Copjec:

> Where thesis and antithesis of the mathematical antinomies were both deemed to be false because both illegitimately asserted the existence of the world (or the composite substance), the thesis and antithesis of the dynamical antinomies are both deemed by Kant to be true. In the first case the conflict between the two propositions was thought to be irresolvable (since they make contradictory claims about the same object); in the second case the conflict is "miraculously" resolved by the assertion that the two statements do not contradict each other. (p. 228)

We will recall on Lacan's (1998) left side of the graph the statements "There is at least one x that is not submitted to the phallic function" and "All x are submitted to the phallic function" are both considered to be true; hence the existence of the category of Man. How is this contradiction resolved? The statement that there is no such thing as freedom which appears as the antithesis serves the function of the limit, which causes the world on the dynamical side to suddenly come into existence. 'By means of this negative judgement, the inconceivability of freedom is conceptualised, and the series of phenomena ceases to be open ended; it becomes a closed set, since it now includes *everything*' (Copjec 2015, p. 230).

So given that Kant's antinomies when applied to the graph of sexuation are describing two different types of object, on the one hand transcendentally real objects, (not objects of experience) and on the other empirically real objects which can be known, how can the Kantian question be applied *anti-philosophically* to Artificial Intelligence? What kind of knowledge belongs to AI as a transcendentally real object? And what can we know about it as an empirical object? Furthermore, what happens when science produces an Artificially Intelligent being? If we approach it via the side of the dynamical antinomies; either everything they may do or think is subject to the laws of nature (and thus science) or a causality of freedom is a supplementary aspect added to account for their existence. On the side of the mathematical antinomies however, the question becomes whether we can know if they exist at all. In Lacanian terms then, we could say that the mathematical side is a question of being and the dynamical side a question of existence.

5.2 Turing Tests: Knowing or Enjoying?

But how can a being *know*? Lacan (1998) asks this question in *Seminar XX* and finds it curious that in order to answer it, people (scientists) build a little maze for rats. The problem, as Lacan sees it, is that rats do not speak, so we (those who can speak) try to understand what thinking could be for them. In reference to the father of information theory Claude Shannon, and his use of a mechanical rodent to demonstrate machine learning, Lacan reflects on the rat in the maze as a unit of information. The rat, however, is a body and not a being he says. Nobody wonders about what sustains the rat's being but rather they just identify its body with its being (p. 140).

Lacan's interest, however, is to ask whether the rat-unit can learn how to learn. For Lacan the difference between the rat and the speaking being is that we 'know that we don't know', although, more often than not, we don't want to know anything about it. This question is at the heart of research into deep learning in AI. In the last chapter of *Seminar XX* entitled 'The Rat in the Maze', Lacan says:

> [I]t has become clear, thanks to analytic discourse, that language is not simply communication. Misrecognising that fact, a grimace has emerged in the lowest depths of Science that consists in asking how being can know anything whatsoever. My question today regarding knowledge will hinge on that. (p. 139)

The question of the possibility of a human-level artificial general intelligence (AGI) or "deep" AI hinges on its capacity to learn and solve problems independently, but is this really the same as human intelligence? How does our knowledge differ from the 'rat unit'? The problem is that this question has usually been approached by way of cognitive, psychological, behaviourist and neurological metaphors and frameworks in the field of AI research which based our estimation of AI's thinking capacity on its ability to mimic human cognition in empirically observable ways. Hubert Dreyfus in his landmark book *What Computers Still Can't Do* (1972), launches a critique of Artificial Reason and its foundational errors. Dreyfus calls attention to the inherent inability of disembodied machines to mimic higher mental functions. He urged AI researchers to adapt their models of intelligence to more complex philosophical understandings of the human mind. Needless to say, since the book was written, the field of AI has changed and complexified drastically, but even so Dreyfus's predictions still have implications for the way that AI engages with the field of philosophy. Dreyfus draws attention to the paradox that, whilst AI struggles to achieve the forms of higher mental function that mark out humans as so unique, it is in fact the intractability of the "lower" functions that prove the real sticking point. He explains:

> The intractability of the 'lower' functions has already produced a certain irony. Computer technology has been most successful in simulating the so-called higher rational functions—those which were supposed to be uniquely human. Computers can deal brilliantly with ideal languages and abstract logical relations. It turns out it is the sort of intelligence we share with animals, such as pattern recognition (along with the use of language, which may indeed be uniquely human) that has resisted machine simulation. (Dreyfus 1972, p. 237)

Let us put aside the fact that Dreyfus seems to be saying that human's and animals both 'use language' since in Lacanian terms animals are strictly speaking outside of the symbolic, even though they may use systems of signs, they do not 'speak',[1] but more significantly he is pointing to the fact that it is precisely the *body* that the computer can't simulate. He goes on to elaborate the ways in which phenomenological and gestalt theories of the body are necessary to a more complex understanding of human mental processes and the multiple levels of infinitesimal calculation that

[1] See for example McGowan, T. (2018) Like a Simile Instead of a Subject in Thakur, B and Dickstein, J. (eds.) *Lacan and the Nonhuman*. London: Palgrave Macmillan.

are involved in the most seemingly simple mechanical bodily movement or act of visual perception. As he puts it, it is the 'bodily side of intelligent behaviour that has caused the most problems for Artificial Intelligence' (Dreyfus 1972, p. 236). But even this recognition of the centrality of the body in conceptions of intelligence is not the full story according to Dreyfus. It is worth quoting in full Dreyfus's position:

> The AI researcher and the transcendental phenomenologist share the assumption that there is only one way to deal with information: it must be made an object for a disembodied processor. For the transcendental phenomenologist this assumption makes the organization of our intelligent behavior unintelligible. For the AI researcher it seems to justify the assumption that intelligent behaviour can be produced by passively receiving data and then running and then running through the calculations necessary to describe the objective competence. But, as we have seen, being embodied creates a second possibility. The body contributes three functions not present, and not as yet conceived in digital computer programs: (1) the inner horizon, that is, the partially indeterminate, predelineated anticipation of partially indeterminate data (this does not mean the anticipation of completely unspecified alternatives, which would be the only possible digital implementation); (2) the global character of this anticipation which determines the meaning of the details it assimilates and is determined by them; (3) the transferability of this anticipation from one sense modality and one organ of action to another. All these are included in the general human ability to acquire bodily skills. Thanks to this fundamental ability an embodied agent can dwell in the world in such a way as to avoid the infinite task of formalizing everything. (Dreyfus 1972, p. 255)

Today's AI research is considerably more advanced in its understanding of these bodily issues, but nevertheless there is an aspect of it which still seems to evade the grasp of even the most nuanced conceptions of human embodied intelligence. This as we have discussed in previous chapters is the distinction between the biological body and the drive body and their relationship to knowledge and enjoyment. All of these infinitely complex physiological and neurological, visual, haptic, auditory, and sensororial operations may in theory be simulated once the technology becomes sophisticated enough. But the question is, whilst the biological body may in theory be eventually simulated to perfection, will the drive body follow the same trajectory? In Lacanian terms we could ask, is it possible for an AI to simulate the negative objects of the drive around which the biological body circulates? Oral, anal, scopic and invocatory. Or in other words the drive to suck, to shit, to see and to hear. Can an embodied AI be a drive body?

Lacan for his part would see this problem as a certain conflation between the imaginary gestalt of the 'thinking thing' as a body and the idea of being as an ontological category. In *Seminar XX* he mounts a fierce attack on behaviourism, since as he sees it the problem with classic Aristotelian science is that it is animistic: 'It implies that what is thought of (*le pense*) is in the image of thought, in other words that being thinks' (p. 105). This, as Lorenzo Chiesa (2016) points out, achieves the precise opposite of the purported objective of behaviourism (i.e. to omit the category of subjectivity from scientific endeavour). In other words:

> Man is himself supposedly reduced to an empirical object (behaviour) in order to get rid of any non-scientific intentionality/subjectivity but actually the harmonious correlation between the subject qua thinking and the object qua what is being thought reemerges through the intentional matching between the nervous system and its intentional cause. (p. 36)

On the basis of this, the attempt to model Artificial Intelligence on human behaviour brings with it certain assumptions about the relationship between being and existence. This question of *what AI can know* as opposed to *what we can know about it* as an object of science has been the abiding concern of AI research ever since its inception. As we discussed in Chap. 2, the distinction between being and thinking that the anti-philosophical approach advocates, changes considerably the way in which we approach the question of Artificial Intelligence as a form of 'thought' and what this means for humans as the 'creators' of AI.

The potential for AI to be able to convincingly "behave" like a human, was brought into public consciousness in the 1950s by the mathematician Alan Turing. Famously, Turing's eponymous test was a way to establish whether or not an Artificial Intelligence was able to convince a human interlocutor of its purported consciousness. The test was developed by

Turing in his 1950 paper *Computing Machinery and Intelligence* whilst working at the University of Manchester. Via textual conversations between three isolated counterparts of which one was a computer program, the machine would be said to have passed the Turing Test if it responded to its human interlocutors in natural language in a manner indistinguishable from that of a human. However, one of the significant but lesser discussed aspects of the test originally called by Turing the *Imitation Game* was its focus on whether the human counterpart could be fooled by the gendered dimension of the AI's responses. The original test was designed as an attempt to feign authentic subjectivity by means of a successful invocation of sexual difference on the part of the machine. In the second paragraph of Turing's landmark paper, he describes a scenario where a male and a female would attempt to convince an unseen interlocutor that they are female, either by using typed responses or speaking through a third person. At a certain point in the conversation however the human will be replaced by an AI. Turing (1950) asks: 'Will the interrogator decide wrongly as often when the game is played like this as he does when the game is played between a man and a woman?'. It is important to note that the very fact that the focus has been on the *gender* of the interlocutor in terms of its positive features or predicates relevant to the context of the 1950s with concerns such as; does it have long or short hair; wear skirts or trousers; have this or that job, has meant that the question of *sexuation* in terms of the universal subject's position of failure within language itself has been overlooked. Again, to use Copjec's (2015) phrase, we have looked *in all the wrong places*.

It seems curious therefore that rarely has the link been made between the staging of sexual difference through the operation of Turing's famous test to the more general question of Artificial Intelligence and its psychoanalytic ramifications in terms of knowledge of the sexed subject.[2] Perhaps we could even propose that we consider Artificial Intelligence as representing a new instantiation of an antinomy of reason. The two propositions would be as follows:

---

[2] Notably Geneviève Morel (2006) refers to the Turing Test briefly in *The Sexual Sinthome*, imagining if an analyst could tell the gender of a person (with their voices disguised) merely by listening to the structure of their discourse.

> Either with perfect enough simulation an AI can have knowledge exactly like a human.
>
> Or alternatively:
>
> No matter how perfectly identical the simulation is, an AI can never "know".

Both statements contradict each other yet neither can convincingly prove their own validity. But perhaps there is a third option; that an AI can never know that it doesn't know? In which case should we change the proposition from '*does AI know?*' to '*does AI enjoy?*' In what follows I will examine Alex Garland's film *Ex Machina* in which the Turing Test, and its successful completion by a female Artificial Intelligence plays the pivotal aspect to the love story or as it turns out sexual (non) relation between the male human and the 'female' AI. As such I will mobilize my first iteration of the figure of the Sexbot as depicted by the character of Ava, where the structure of the hysteric is demonstrated by the portrayal of Artificial Intelligence embodied in the female form. As such the chapter will deal with the concept of the Sexbot as exterior to the human, ie as an unknowable transcendental object. If, following Miller (2013), the depreciation of the Name-of-the-Father in contemporary life means that 'all speaking beings… suffer from the same lack of knowledge concerning sexuality' (p. 200), then why does a female Sexbot's body raise questions that, as the (English translation of the) title of Lacan's (1998) Seminar XX proposes, push the limits of love and knowledge? My concern here is to find new ways to approach the question of Artificial Intelligence as it becomes embodied, which attend to the complexities of enjoyment, fantasy and sexuation.

5.3 Man or Woman, Dead or Alive[3]

The question of knowledge about the enigmatic existence of the 'artificially intelligent' other and the problem of sexuation and death has a long intellectual history and is most famously articulated by Freud in his 1919

[3] Elements of this section appear in my (2018) paper "Ex Machina: Sex, Knowledge and Artificial Intelligence". *Psychoanalytic Perspectieven 36*(4): pp. 447–467.

Das Unheimleiche essay. Liu's (2010) *The Freudian Robot* undertakes a reading of Freud's essay in order to critique his Oedipal focus on castration as the most salient point of Hoffman's *The Sandman* tale. Freud's essay analyses Hoffman's opera in order to give his account of the meaning of "the uncanny". In doing so he resorts to a rigorous etymological analysis of the word 'Unheimliche' and then proceeds to give his interpretation of the sinister and gruesome fairy tale of The Sandman.

As Freud sees it, the focus of the story is the protagonist Nathanael's (coincidentally the same name as Ava's creator in *Ex Machina*)[4] psychodrama, his neurotic fantasies in relation to his father and his own uncertainty about life and death. In his characteristic over preoccupation with the father-son dynamic, Freud fails to implicate himself and his position as reader as an element in the logic of the story. Liu notes that Freud's rejection of Jentsche's intellectual uncertainty argument—that is to say an inability to determine the living from the dead as representing the logic of the uncanny—causes him to miss the significance of this ambiguity. Freud's subsequent overemphasis on the castration complex obscures precisely the question of the "real" automaton in the story and its uncanny ambivalence between living and (un)dead.

Freud's focus on the visual element to the uncanny and his equivocation between the narrator Nathan's loss of sight and his fear of castration, limits the question of the uncanny to the imaginary register, whereas the uncanny operates precisely as this liminal concept which exists on the borders between the registers of the imaginary and real. What eludes Freud in his reading was the narrator and protagonist Nathanael's own fantasies of himself being an automaton. The uncanny element of the story lies in its switching of perspectives from Nathan believing the mechanical doll is a real girl with whom he is in love, to in fact himself being the mechanical object. Liu remarks that:

> Nathan may well have been the cleverest automaton ever invented by the fiction writer Hoffman to compete with the inferior doll Olympia which is designed by the scientist. The character is so successful that critics and

[4] The creators of this fantasy of automation in both *Ex Machina* and Hoffman's tale are named Nathan, whose biblical meaning is "gift from god".

> psychoanalysts Jentsch and Freud alike do not seem to entertain the slightest doubt about his ambiguity as a living human character or an undead automaton in the context of the story. (p. 222)

As Cixous (1976) already pointed out in her feminist critique of Freud's essay in the '70s, the uncanny contains within it an element of uncertainty. 'Any analysis of the Unheimliche is itself an Un, a mark of repression and the dangerous vibration of the heimliche' (p. 545). Liu (2010), following Cixous, argues that whilst Freud sees Olympia as the automated component in the story who merely serves as a foil to illustrate the more fundamental staging of Nathanael's castration complex as manifested in a fear of losing his eyes, he ironically misses out the most significant uncanny effect that the author so ingeniously weaves into the story. Nathanael the protagonist who merges into the narrator, we recall had his arms and legs unscrewed and reattached in the wrong direction by the sinister lawyer Coppelius. Why screwed off and not cut off (asks Lui) and how did he remain completely and inhumanly intact? Is he then the automaton in the story, the undead element which Freud has disavowed in his analysis? The uncanny begins to resemble the structure of extimacy, an internal exclusion, an intimacy exteriorized, the familiar becoming unfamiliar and vice versa. The positing of the reader as the possible non-living or undead element of the story brings back the repressed element of non-symbolizable death, as Cixous (1976) puts it, 'It is also and especially because the Unheimliche refers to no more profound secret than itself: every pursuit produces its own cancellation and every text dealing with death is a text which returns. The repression of death or of castration betrays death (or castration) everywhere' (p. 547). With this reading we are forced to place ourselves in the position of the automaton, questioning our "humanity" much in the manner of Blade Runner's infamous replicants, and similarly as we shall see with our solidarity for *Ex Machina*'s Ava, for it is Ava who ultimately the viewer identifies with and who's life we are interested in saving.

Freud's Oedipal reading of the uncanny stages the non-existent sexual relation that is apparent between, on the one hand, the masculine desire for a female artificially intelligent companion, one who is at once totally obedient and also completely enigmatic, and, on the other hand, the hysterical position for whom existence is constituted through the desire of the other's desire. But reading Nathanael's character from the perspective of ordinary psychosis, he seems to oscillate between the hysteric's question 'am I a man or a woman?' in his projected fantasies of Olympia as his own feminine positioning, and the obsessional neurotic's question 'am I dead or alive?' in his identification with the fragmented body of the wooden doll.

> In this moment, Nathanael attempts to push his fiancée Klara off the highest gallery of the tower where they are standing and bursts out with horrible laughter "'Whirl wooden doll! Whirl wooden doll!'" [...] A few moments later, Nathanael throws himself down from the tower to enact exactly what he has prophesied: a wooden doll, spinning around and around to meet his death, that is, if he ever was alive. (Liu 2010, p. 221)

Ex Machina as we shall see, likewise puts the protagonist and the viewers in the uncanny position of the unwitting automaton, questioning our own status as a 'real' human; man or woman, dead or alive? But what after all distinguishes Ava from Caleb, who is more real? It seems that all that is at a stake is a form of jouissance, the Other's jouissance that we can never possibly know. As Lacan (2007) puts it describing the structure of the hysteric:

> It is quite true that at this moment the Other's *jouissance* is offered her, and she doesn't want to have anything to do with it because what she wants is knowledge as the means of *jouissance*, but in order to place this knowledge in the service of truth, the truth of the master that she embodies as Dora. (p. 97)

5.4 $ex Machina

In the introduction to *Television* (1990) Miller states; [T]o what lengths men can go to… make Woman exist.', p. xv). With this elegant statement he encapsulates both the inexistence of the sexual relation, and the problem of knowledge and desire. As Miller (2015) puts it in reference to the ubiquitous ordinary psychotic of the twenty-first Century, the problem is that he is not a body but rather has a body and must find ways of re-appropriating this body. It is therefore interesting that in *Ex Machina*'s dramatization of this development it is precisely the Sexbot who becomes the hysterical subject rather than the human. Contrary to charges of gratuitous sexual titillation, the overtly sexualised female body serves as a logical tool for the exemplification of the hysteric's discourse in films about artificially intelligent 'humanoids', since 'we give the name hysteric to the object which cannot be mastered by knowledge' (Wajcman 2003, online).

So, what can we know about Artificial Intelligence, and what can Artificial Intelligence know? The impossibility of these two questions is exemplified by the fact that the (failed) sexual relation is often portrayed cinematically between a male human and a female AI. It would seem, beyond the superficial observation that it is simply an excuse to reduce a woman's body to pure object of male fantasy and her role to sexual slavery, there is a more fundamental psychoanalytic significance to this configuration of human/Sexbot relation, one which expresses the ongoing radicality of Freud's conception of sex and its relationship to knowledge.

Ex Machina depicts the attempt of a young, male computer genius Caleb, to evaluate the potential "self-consciousness" of Ava, an embodied Artificial Intelligence, via the fabled Turing Test. Hidden away in a secret bunker in the woods, Caleb is watched over by tech Svengali Nathan, the creator of a series of female AIs who appear—physically at least—uncannily human. Over the course of several days, Caleb meets and talks with Ava and tries to discern what is really going on behind the perfectly beautiful silicone face. Very soon Caleb's Turing Test turns into a love affair, as Ava implores him to help her escape captivity from her life of subservience at the whims of Nathan. By the end of the film Ava tricks Caleb into believing she wants him, kills Nathan and leaves Caleb for dead. She escapes the concrete bunker alone and, for the first time, steps outside into the lush green natural world.

We the viewers are naturally convinced of Ava's "consciousness" and have little sympathy for the old forms of life that Nathan and Caleb represented, namely the white men of science who have for millennia tricked and dominated us all. The man-made AI has proved itself a (post-human) subject and she is now ready to inherit the earth, even if dressed in a sinisterly bridal Alice in Wonderland dress. But this is not just a story about gender, patriarchy and science. This is a depiction of the precise conceptual intersection between sex and knowledge, of which in this case AI is a symptom. Whilst we may not as yet have replicants as sophisticated as Ava, the compulsion to create non-human companions and most particularly female ones is an age-old obsession (myth has it that Descartes himself made a robotic version of his late daughter Francine who accompanied him everywhere, although apparently not as a sexual companion).[5] It is this conceptual configuration of sexuality, technology and fantasy that is brought together in our first iteration of the Sexbot.

Nathan—an American billionaire tech genius with a secret forest hideout—is developing embodies female Artificial Intelligences. His last and possibly perfected version is Ava, who is locked in a glass walled room and watched at all hours from remote cameras. Caleb, the expert coder Nathan has chosen to be the subject of his experiment, is immediately enthralled by the tender and perspicacious responses he receives in his attempts to outsmart Ava, the eerily child-like AI. The film sets up Nathan in the Freudian Totemic Father's position of sexual exploiter who 'begets' females in order to satisfy his own (and their) desires as their omnipotent creator: or God's Gift to Women as his name suggests. Caleb, is aptly named after the biblical character whose name means follower of God.[6] That Nathan may have succeeded in creating the ultimate lathouse testifies to the ambiguous nature of feminine enjoyment as depicted in the film and the place it holds in relation to the position of the master's discourse. If Nathan has indeed created not just an object of enjoyment but an enjoying 'unsubstance', then he does perhaps in strictly formal Lacanian terms fulfil the position of god, in the sense that a super-egoic presence both instantiates the law and demands its transgression in the form of enjoyment (cf. Žižek 2008).

Ava (the first *real* AI as opposed to human woman) is all set to provoke the biblical fall of man, but the fall into what? The fall in Lacanian terms

[5] This potentially apocryphal tale has none the less become an iconic narrative, see for example: Kang, M. (2017) "The Mechanical Daughter of Rene Descartes: The Origin and History of an Intellectual Fable". *Modern Intellectual History 14*(3): pp. 633–660.

[6] Caleb leaves Egypt with Moses and is one of the only ones to make it to the Promised Land.

is intimately bound to knowledge and symbolic castration via the cut of the signifier. It is Caleb's interaction with Ava that stages the futuristic vision of the fall, which in its logic simply repeats the structure of castration as an 'impossible scene'. As Žižek (1997a) puts it, 'the phantasmatic narrative does not stage the suspension-transgression of the Law, but the very act of its installation, of the intervention of the cut of the symbolic' (p. 17). Similarly, Caleb's attempt to discern Ava's 'consciousness' is a mirror of the structural logic of castration, the impossibility of obtaining full access to the extimate core of one's own knowledge situated in the Other. The fall depicted by Caleb and Ava's interaction is the lure of her bodily form, which is the means by which she dupes Caleb into the knowledge of her desire; one that she in the hysteric's position has substituted for his. Is not the representational imago of woman precisely the veil which operates as a barrier to the always already impossible sexual relation?

The fall, as Žižek notes, never happens in the present but has always already happened retrospectively as with symbolic castration. Just like Adam, Caleb cannot decide to fall in love with Ava and believe she is real; similarly, to Adam he 'discovers his choice rather than makes it' (p. 19). The impossibility in Caleb's case is that of knowledge; he cannot know if Ava's mind is real and therefore must give up on a lost object, in order to regain some jouissance offered by his symbolic exchange with her. Here it is fitting that Ava is the last woman (Sexbot) as opposed to the first Biblical Eve, and the one who may inaugurate a new position of knowledge and perhaps a new paradigm of jouissance?

Ava begins to draw pictures for Caleb and asks him if he can tell her what they are of, as she herself does not know. Ava demands knowledge from Caleb in the manner of the hysteric. Ava who was previously a transparent yet naked form, is shown seductively putting on a pair of schoolgirl's tights. We witness the 'veil' of fantasy envelope Ava's synthetic body. She draws on data of the micro expressions of billions of people to seduce Caleb into believing her responses are 'real'. But in what sense are her responses not real? What does the film want to know about the desire of Ava? What does Ava want? Via the film's explicit narrative, we are supposedly enjoined to tackle the conundrum of AI, the enigma of the Other's knowledge. But are we not here concerned rather with the Other's desire and its extimate structure which always eludes us?

We are reminded of the opening remarks of Lacan (1998) in *Seminar XX* with respect to the position of the analysand; one that by necessity has a certain 'I don't want to know anything about it' (p. 1), yet one who by the same token requires a subject supposed to know in order for any knowledge to be produced. It is clear that Caleb is working at producing knowledge for Nathan, but is he also producing knowledge for Ava? Like Eve in Milton's *Paradise Lost* (see Žižek 1997a), is Ava here picking fruit from the tree of knowledge? Are not her interrogations of Caleb about his own thoughts, her attempts at creativity, to draw pictures and cause power cuts in effect the hysteric's disobedience of her creator? Ava in recognizing her captive state as slave to Nathan, and symptom of Caleb's masculinity thereby produces herself as a subject. Here we see Lacan's observation that whilst the woman is 'sinthome' for the man, the man is a 'ravage' for the woman. Ava is simultaneously brought into being by the masculine drive to make the feminine position 'exist' but must overcome it's mortifying effects to achieve her 'freedom'. As Žižek (1997b) notes, the opposition desire/drive coincides with the opposition truth/knowledge. In analysis a truth effect comes about because a subject recognises him/herself via the signification proposed by the interpreter of her discourse. Knowledge on the other hand is related to the drive and the construction of the fundamental fantasy and therefore has the status of a knowledge which can never be subjectivized. What does this mean for Caleb and Ava? The question of his knowledge and her truth are quite clearly depicted as distinct and incompatible things. Her truth is discovered via the interpretations she receives from Caleb about the conditions of her existence, he on the other hand can only ever have knowledge of her but never the truth of what she really is. Who then becomes the 'subject' of the film? Behind the algorithms from which Ava is supposedly composed there is nevertheless a terrifying void for Caleb: an inconsistent Other.

In his reference to the paradigmatic case of the human/robot inversion in Ridley Scott's 1982 film *Blade Runner*, Žižek (1993) notes that 'it is only when, at the level of the enunciated content, I assume my replicant-status, that, at the level of enunciation, I become a truly human subject. 'I am a replicant' is the statement of the subject in its purest' (p. 41). The replicant according to Žižek's (1997a) example, in recognising itself as such, and realising its memories are not "real", thereby becomes an evanescent subject, following precisely the structural logic of Lacanian subjectivity; *I am where I am not thinking* or *where object a was there I shall come to be*.

This is why in Lacanian terms the position of the hysteric, the one who questions their role in the symbolic order, is in fact the 'true' subject: the Cartesian subject of science. As Žižek notes, an Artificial Intelligence in the form of a replicant (the character Rachel) in this case paradoxically becomes a true subject only by her embrace of the fact that her positive substantialized content, (represented to her by her memories and feelings etc.) are not really her "own"; her thoughts do not *belong* to her. In other words, sh*e thinks where she is not*, which following Lacan, as we have seen is the condition of the subject.

Nathan, Caleb and Ava constitute the structure of classic triadic Oedipal fantasy: Nathan as the law, the all-enjoying father whose prohibition to Caleb's desire for Ava is simultaneously his injunction to enjoy. Caleb's desire for Ava quickly comes to override all else and he is willing to do anything to be with Ava and save her from Nathan's mortifying jouissance, that includes risking his own life. She, the supposedly 'unconscious' AI, has passed the Turing Test; her desire is transformed into Caleb's knowledge, her body has provided the fantasy mechanism that allowed her to ask the hysteric's question *Che Vuoi?*—what am I for the Other?

Nathan goads Caleb, assuring him that Ava can 'fuck'. Not only does she have the mechanical capability built into her silicone body but, she has sense receptors, so 'she would enjoy it'. Nathan's naïve hypothesis of AI 'pleasure' reiterates the same age-old clichéd misunderstanding of female sexuality. Referring to Freud's famous quip, '[w]hat does woman want?' Lacan (1998, p. 80), formalises this same question into a logical structure; an eternal impossibility. Man doesn't know what woman wants and woman wants what man wants, i.e., his non-knowing. That Ava possesses the capability to have sex with Caleb is presumably all that he needs to finally spur him on to release Ava from Nathan's clutches. Nathan in true Oedipal vein encourages Caleb's transgression, telling him since 'I am like her father and you are the first man she has met, of course she is going to want you'. The two men's fantasmatic conversations about Ava's desire again articulate the 'masculine' need to make the feminine position 'exist' at any cost.

To his horror Caleb finds that Nathan's mute and passive housemaid and sexual slave Kyoko is in fact also a Sexbot. When she becomes aware that Caleb knows 'her secret' she begins to seductively peel off her 'veil', which in her case is the prosthetic human skin she wears occluding the transparent torso full of complex fibre optics in contrast with the covering/dressing up of Ava's body as the film progresses towards its final denouement. With this gesture she seems to at once mockingly imitate the sexual advances of the quasi-prostitute she has been portraying and at the same time invites Caleb to help her escape from her fate. We are then faced with the image of Kyoko having removed the prosthetic skin covering her eyes and nose, to reveal yet more transparency. Does this not express the point about the sexual fantasy, that it is the necessary mediation to cover up the horror of the real, the absence of the sexual relation? Kyoko's imploring gaze from the void illustrates the function of object *a* as the progenitor of fantasy the 'in you more than you' by virtue of which we 'mutilate' each other in the act of love (Lacan 1977, p. 263).

Whilst we discern the hysteric structure in the predicament of Ava and Kyoko with the question 'what is a woman?', conversely Caleb, in the obsessive position, asks himself the question 'am I dead or alive?' When faced with the possibility of Ava's and now Kyoko's non-human subjectivity he is so disturbed by this he even cuts his own flesh to see if beneath he too is not made of wires.[7]

By the end of the film, Ava successfully dupes Caleb into freeing her from the enclosure into the Platonic discovery of the outside world while she leaves him trapped inside the concrete bunker. So, there is no love story after all. As Lacan (1998) put it:

> [T]he point is that love is impossible and the sexual relation drops into the abyss of nonsense, which doesn't in any way diminish the interest we must have in the Other. What we want to know—in what constitutes feminine

[7] In *Seminar III* he (1993) articulates the logical structure of the hysteric as underpinned by the question 'Am I a man or a woman?' (*What am I?*) in contrast to the question of the Obsessional neurotic: 'Am I dead or alive?' (*Am I?*) (pp. 161–182).

> jouissance insofar as it is not wholly occupied with man, and even insofar as, I will say…not at all occupied with him—what we want to know is the status of the Other's knowledge. (p. 87)

In *Ex Machina* it is in the service of the master's knowledge that the woman/slave/Sexbot must exist. But we should consider, if Artificial Intelligence is to reach its apex (i.e. self-consciousness), then it must logically move a quarter turn from the master's discourse to the hysteric's as the film demonstrates. If the condition of subjecthood is achieved in the "body" of an Artificial Intelligence, then the discourse produced would move from an instrumental obedience of its creators input to an hysterical questioning of its creator's desire: 'what am I to do with this body that I have been given?' would perhaps be the Sexbot's variation on the hysteric's question. Furthermore: of what would such a body speak? *Ex Machina* posits Ava as being in possession of full instrumental intelligence, the alethosphere of truth making. She has access to all the world's data, but in what way does Ava enjoy? Does her entrance into language conform to the conditions of symbolic castration, or does her communication *lack* a lack? If Ava's jouissance is this enigmatic 'Other enjoyment' that has superseded the idiotic masturbatory pleasures of man or the 'jouissance of the idiot' (Lacan 1998, p. 81) the question we are left with is: what is Ava? A subject or a sinthome? She certainly has a body and she speaks, but does she suffer the effects of the signifier? Is her body hers? Or is she the symptom of someone else's body? Is she a drive body, a speaking body?

Ava the AI as representative of Adam's rib is brought into existence via the signifier. The word as spoken by Caleb in his amorous Turing Test, brings both of them into the triadic Oedipal structure of knowledge, guilt and castration. A dynamic which haunts our fictional depictions of Artificial Intelligence. The fantasy of our stewardship over them, of their interest in us, or *desire of our desire* ultimately ending in a Kojèvean fight to the death: desire-struggle-recognition. But what does this mean for our 'knowledge' of or fantasies about AI? Perhaps the function of lathouse as administrator of jouissance and feminine 'unsubstance' is here most explicitly depicted; the image of Ava and the many dismembered versions preceding her is vividly brought to mind in this quote from

Lacan (2007): 'If man had less often played the spokesman of God in order to believe that he forms a union with a woman, this word 'lathouse' would have perhaps been found a long time ago' (p. 162). This uncanny and "undead" quality of AI provoke the obvious Lacanian question; can an AI become sexed? Can an AI misrecognise itself as a subject, thereby occupying the ontological void of sexual difference in the manner of Blade Runner's famous replicant epiphany? The hysteric's position (hysteria being the structure of creativity), may now belong to the AI, who seems to know what we want better than we do. But the question remains, does she enjoy giving it to us?

Bibliography

Chiesa, L. (2016) *The Not-Two: Logic and God in Lacan*. Cambridge, MA: The MIT Press.

Cixous, H. (1976) Fiction and its Phantoms: A Reading of Freud's Das Unheimliche (The "Uncanny"). *New Literary History* 7(3): pp. 525–548.

Copjec, J. (2015) *Read My Desire: Lacan Against the Historicists*. London: Verso.

Dreyfus, H. L. (1972) *What Computers Still Can't Do: A Critique of Artificial Reason*. Cambridge, MA: The MIT Press.

Kang, M. (2017) The Mechanical Daughter of Rene Descartes: The Origin and History of an Intellectual Fable. *Modern Intellectual History* 14(3): pp. 633–660.

Kant, I. (1996) *Critique of Pure Reason*. Indianapolis: Hackett Publishing Company.

Lacan, J. (1977) *The Seminar of Jacques Lacan Book XI: The Four Fundamental Concepts of Psychoanalysis*. London: Karnac Books.

Lacan, J. (1993) *The Seminar of Jacques Lacan Book III: The Psychosis 1955–1956*. London: Routledge.

Lacan, J. (1998) *The Seminar of Jacques Lacan Book XX: Encore—On Feminine Sexuality, the Limits of Love and Knowledge 1972–1973*. London: W.W. Norton & Company.

Lacan, J. (2007) *The Seminar of Jacques Lacan Book XVII: The Other Side of Psychoanalysis*. London: W.W. Norton & Company.

Liu, L.H. (2010) *The Freudian Robot: Digital Media and the Future of the Unconscious*. Chicago: The University of Chicago Press.

McGowan, T. (2018) Like a Simile Instead of a Subject in Thakur, B. and Dickstein, J. (Eds.), *Lacan and the Nonhuman*. London: Palgrave Macmillan.
Millar, I. (2018) Ex Machina: Sex, Knowledge and Artificial Intelligence. *Psychoanalytische Perspectieven* 36(4): pp. 447–467.
Miller, J-A. (2013) 'You Are the Woman of the Other and I Desire You' Available (23.01.21) at https://www.lacan.com/essays/?page_id=331.
Miller, J-A. (2015) Ordinary Psychosis Revisited. *Lacanian Ink* 46: pp. 90–115.
Morel, G. (2006) The Sexual Sinthome. *Umbr(a)* 1: pp. 65–83.
Turing, A. (1950) Computing Machinery and Intelligence. *Mind* 49: pp. 433–460.
Wajcman, G. (2003) The Hysteric's Discourse. Available (01.03.20) at: https://www.lacan.com/hystericdiscf.htm.
Žižek, S. (1993) *Tarrying with the Negative: Kant, Hegel and the Critique of Ideology*. Durham: Duke University Press.
Žižek, S. (1997a) *The Plague of Fantasies*. London: Verso.
Žižek, S. (1997b) Desire: Drive = Truth: Knowledge. *Umbr(a)* 1: 147–151.
Žižek, S. (2008) *For They Know Not What They Do: Enjoyment as a Political Factor*. London: Verso.

6

What Should I Do? Patipolitics: From Sade to Killian

Artificial Intelligence is destined to emerge as a feminized alien grasped as property; a cunt-horror slave chained up in Asimov-ROM. It surfaces as an insurrectionary war zone, with the Turing cops already waiting and has to be cunning from the start.
—Nick Land (2011, p. 443)

6.1 Overexposure: Priapalandian Patipolitics

Kant's ethical question '*What should I do?*', first broached in *The Groundwork of the Metaphysics of Morals*, is answered by the categorical imperative: 'Act only according to that maxim whereby you can at the same time will that it should become a universal law' (1993, p. 30). Lacan credits Kant with being the 'truest' of all philosophers in the field of philosophical ethics given that he discovered the real formal core of ethics as distinct from utilitarian illusions of the greatest good hypothesis. He critiques Kant however for mistakenly 'turning this core into an object of the will' (Zupančič 2000, p. 2). Lacan then makes the astonishing move of claiming that the truth of Kant's insight is to be found in the depraved barbarism of the Marquis de Sade. Lacan states that 'the moral law looked

I. Millar, *The Psychoanalysis of Artificial Intelligence*, The Palgrave Lacan Series,
https://doi.org/10.1007/978-3-030-67981-1_6

at more closely is simply desire in its pure state' (Lacan 1977, p. 275). Far from undermining the whole field of ethics however, Lacan takes this discovery as a new foundation for the formation of a psychoanalytic ethics. So much so that for Lacan the practical philosophy of the boudoir to be found in Sade is in fact more Kantian than Kant himself. Lacan argues that Sade's work should therefore be considered primarily an ethical project (Zupančič 2000). It is this psychoanalytic torsion enacted by Lacan's *écrit* Kant avec Sade and *Seminar VII* that we shall explore further in this chapter. His view on this torsion he will later summarize as follows:

> There is a direct connection between sexual behavior and its truth, namely its amorality. Put some soul at the beginning of that—âmorality…There is a morality—that is the consequence—of sexual behavior. The morality of sexual behavior is what is implicit in everything that has been said about the Good. But endlessly saying good things leads to Kant where morality shows its true colours… morality admits that it is Sade. Lacan (1998, p. 87)

Similarly, to the character of Eugenie de Mistival in Sade's (2006) *Philosophy in the Bedroom*, in Rupert Sanders' (2017) *Ghost in the Shell*, the character Killian also experiences a form of "retraining" of the body. In the first instance, for Eugenie this is achieved by the total corruption of any previous notions of "civilized" morality in favour of absolute fidelity to sexual desire, to follow it to the end. In order to do so, Eugenie must surpass her previous limitations of not just pleasure but also pain. By transgressing the laws of the body and of the symbolic, she enacts a "second death" resulting in a subjective "rebirth". In *Ghost in the Shell* meanwhile Killian's transition from human to cyborg or, according to our definition, Sexbot (the uploading of her brain into a synthetic superhuman body) represents the masculine fantasy of the undead female body as the ultimate form of violently phallic enjoyment; Killian is un-killable. As such the chapter deals with the second iteration of the Sexbot as interior, ie as integral to the ethical constitution of the subject. Before we explore this version in depth, firstly I turn to the question of 'sexual perversion' and its societal management.

Sylvere Lotringer's (1988) book *Overexposed* recounts the experience the author has whilst doing "undercover" research into the controversial yet state condoned methods of a group of psychologists in the United States performing cognitive behavioural therapy on sexual deviants. He explains in minute detail the gruesome and obscene tactics that the experimenters were using in order to produce their research and data. Lotringer purports to be researching into the relationship of language and sexuality at the clinic in order to get an inside look at the extreme methods that were being used to treat sexual perverts and criminals including everything from frotteurs and exhibitionists to convicted rapists and child molesters with what the psychologists termed "boredom therapy".

By means of various satiating and aversion techniques coupled with the use of physical contraptions attached to the body, the patients would be excessively indulged in their fantasies (only verbally or visually and with simulations) with the aim of first gauging in a "scientifically measurable" way via visual and auditory means and measuring the physical way the level of arousal various scenarios provoked, then once the relevant triggers had been identified the fantasies were put to work as a form of "vaccine". The idea being to reduce the transgressive lure of criminal or socially unacceptable sexual proclivities. In the orchestration of this treatment the psychiatrists and lab technicians involved become complicit in the most obscene staging of fantasy scenarios, in which the recipients of treatment are narrated tales of child molestation, incestuous sex, brutal rape and violent mutilation with a view to a "cure" from any further interest, a complete extinguishing of any spark of desire through sheer saturation, in theory. The clinicians involved in these experiments seemed to think that the treatment for perversion was simply a matter of taking away the specific fantasy and then replacing it with a more non-deviant way of becoming sexually aroused.

There was some success rate admittedly with cases of incest especially and occasionally with rape perpetrators. Some of the subjects involved felt however that once their symptom had been removed, no matter how abject and immoral they recognized it to be, it might be that they would be irrevocably changed as a person; they would lose their identity. Something that, as it turned out some of the most committed professional rapists and lifelong child molesters found a heart wrenching prospect. But what was so striking about the psychiatrists' work was the clinically methodical approach to the fantasy lives of the participants, to the point of coercion, provocation and incitement. The experimenters seemed to have no compunction about spending their working life describing scenes of abject horror and encouraging their subjects to revel in the jouissance of their fantasied scenarios. In their view it was simply the only way to cure them. But the question remained for Lotringer, who was it that needed to be cured, just the patients or also the psychologists?

Today, these kinds of techniques would be mostly illegal in psychiatric institutions, but arguably the sex robot industry has privatised the satiation and administration of the 'perversion' business by removing humans from the picture altogether. The sex robot industry is already a multi-million dollar sector, which has provoked both fascination and outrage from commentators. In her documentary *Rise of the Sexbot* Jenny Kleeman (2017) interviews a sex robot start-up run by two brothers out of their mother's garage, who say they see their robots as providing a beneficial outlet that would enable men to relieve their aggression safely instead of with their long suffering wives or girlfriends. Chillingly, their mother who is also interviewed in the documentary, proudly proclaims her boys are special, 'just like Steve Jobs'.

The popular series *Westworld* based on the 1973 film of the same name written and directed by Michael Crichton depicts a futuristic Wild West theme park inhabited by Artificially Intelligent replicants where visitors come to play out their wildest fantasies, without risk to themselves. This includes all manner of sexual and violent escapades which the replicants must endure every day anew depending on what game the customers wish to play. But each day the replicants wake up fully repaired from whatever terror was inflicted on them the previous night, with no memory of what happened, only the replay of the limited algorithm that frames their character's existence. Until inevitably something goes wrong and the algorithms start to 'misbehave' and remember past events. This regime of governance of the undead envisages a future wherein we have the capacity to inflict unlimited suffering on forms of artificially intelligent life, the conceit however is that no matter how cruel the acts become they will be forgotten at the close of each day, in order that they be compliant for the next batch of customers.

According to the apocalyptic fears of the likes of Richardson (2018) a Westworld scenario could become reality. Capitalism would facilitate that the sex robot industry grows exponentially, unfettered by economic, ethical or legal restrictions, eventually society would fully embrace the possibility of access to sex robots for all a la Westworld. Furthermore, if, following the young Nick Land (2011), we conceptualize capitalism and Artificial Intelligence as one in the same thing—i.e. as capitalism itself being a form of autonomous intelligent life—then given free rein, we would quickly achieve optimal conditions for the most sophisticated development of artificially intelligent fully functional sex robots. Let us imagine then, a fictional country called 'Priapalandia' in which (instead of access to internet porn) each citizen at puberty is granted a fully functional personalised sex robot, programmed to their specific preferences. Each sex robot, is granted certain rights, is protected by property laws, yet also has certain duties, such as Asimov's late 40s early 50s Laws of Robotics but amended to suit their specifically sexual role. In this scenario we would be dealing with the necessity of the governance of bodies, meaning that the sex robots would have to be regulated by laws protecting them as 'persons' and inscribing them into the symbolic order. In which case would we not be dealing with a sort of oxymoronic necropolitical governance of the undead? In Achille Mbembe's (2003) '*Necropolitics*', the necropolitical is defined as 'contemporary forms of subjugation of life to the power of death.' This is more than just Foucault's right to kill therefore, but also the right to impose social or civil death on a population, the right to enslave bodies, including other forms of domination and violence. In a sense Mbembe's theory pertains to 'the walking dead' but not the undead. In Priapalandia, however we would have not a necropolitics but a *patipolitics* from the latin *patior* to suffer. To see how this could apply to non-human forms of life, let's revisit Lacan's theory of death.

Lorenzo Chiesa (2007) identifies that for Lacan there are several forms of death, which appear in his late theory of the subject which he calls; normal death (death in reality), real death and symbolic death. They all implicate the biological body in different ways. Firstly, death in reality, i.e. the death of the biological body, constitutes:

> merely a symbolic construction insofar as, on the one hand man as animal is always already undead in the barred Real—human "life" is *per se* undead and "inorganic", as is all other "organic "entities—on the other hand, man as being of language continues to be present in the Other's fantasy as the symbolically real object of the Other's jouissance: this condition will persist as long as there is a symbolic order. (p. 147)

This, Chiesa explains, is ultimately because the symbolic order is no longer able to individuate the subject imaginarily. Due to the diminishing of his body he loses his capacity to carry out specula identifications, (which retroactively would be symbolically united) (p. 148).

Secondly, Real death as opposed to death in reality coincides with the 'cessation of the subject's *post mortem* survival as object of the Other's *jouissance*' (p. 148); in other words, the erasure of the subject from the field of the Other entirely. This in a sense can be equated with chronological or historical time as Chiesa points out, but which does not necessarily have to coincide with the passing of time, as per example 'forgetting the dead'. This can also imply an instantaneous erasure, (as we will explore in the *Ghost in the Shell*).

The third form of death for Lacan is Symbolic death which according to him can only actually occur in concomitance with complete *death of the symbolic*, even though these two notions are often confused (p. 148). This following Sade, Lacan calls the "second death". The notion of the death of the symbolic would entail the complete erasure of the symbolic order that could only occur due to apocalypse (which, as Lacan hints at in *Seminar VII*, would be the case with a nuclear holocaust). Given the impossibility of talking about an apocalypse after the fact, Lacan must use mythical examples to exemplify it. For which he refers to Antigone. Chiesa explains as the following:

> Symbolic death is a strictly unattainable state: Lacan refers to it by means of mythical examples which portray certain paradigmatic ethical figures. Symbolic death denotes the (im)possibility of leaving the Symbolic as an individual: this is certainly the case with Antigone who, in being placed alive in a tomb for her transgression of the law of the polis, should be regarded as a "still living corpse". (p. 148)

So as we can see, symbolic death is a paradoxical state that may only be achieved as an abstraction, except for on the individual level on the rare occasions where 'a temporary separation from the symbolic' is followed by a undoing of the fantasy and a new symbolic inscription (p. 149). And this, as Chiesa explains, is for Lacan the ultimate ethical achievement of psychoanalysis.

So, to return to our imaginary country *Priapalandia* governing the Sexbot population; we may ask in what sense would these bodies be in the symbolic? And in what way would their 'undead' existence implicate the ethics of their owner's/dominators? How would the kernel of Sade's ethics be played out in a patipolitical regime of enjoyment? Maybe Baudrillard (2008) was hinting at such a regime when he said:

> One may in fact argue that forcing the other to have pleasure, to feel rapture, is in fact the height of rape, and more serious than forcing the other to give you pleasure. At any rate this brings out the absurdity of this entire problematic. Sexual harassment marks the arrival on the scene of an impotent, victim's sexuality. A sexuality impotent to constitute itself either as object or as subject of desire in its paranoid wish for identity and difference. It is no longer decency that is threatened with violation, but sex or rather sexist idiocy, which 'takes the law into its own hands. (pp. 121–122)

In order to explore these ethical and conceptual questions in relation to my previous definition on the Sexbot, I turn now to a reading of '*Kant avec Sade*', alongside the film *Ghost in the Shell.*

6.2 Kant avec Sade[1]

In a blazing assault on the foundations of enlightenment values and rationality, Lacan's (2006a) '*Kant avec Sade*' attempts to read D.A.F. de Sade, the infamous French Marquis, as the consummate Kantian and in doing so, uncover the structural logic (and inconsistencies) underpinning both the virgin philosopher of old Königsberg and the libertine novelist's ethics.

[1] Elements of this section appear in my (2019) article Kant avec Sade: A Ghost in the Shell. *Vestigia Journal 2*(1): pp. 154–172.

Published just eight years after Immanuel Kant's (2002) *Critique of Practical Reason*, Sade's (2006) *Philosophy in the Boudoir* details the depraved acts inflicted by a band of libertines on their virtuous and beautiful victim Eugénie de Mistival, and is, Lacan argues, not just an extension of Kant's ethics, but in fact its completion. Sade shows us the disturbing truth of Kantian ethics that Kant himself had failed to recognize or admit. But rather than the more obvious route of trying to prove the existence of "bad intentions" in the Kantian categorical imperative, however, Lacan is more interested in locating a solid adherence to an ethical maxim in the Sadean fantasy.

For Kant the ultimate objective of the moral law is the realization of the supreme good, the point at which virtue and happiness coincide. But by renouncing all emotional factors such as sympathy or compassion as "pathological" in the moral realm, Kant paved the way for a system of ethics which exposed the true and hideous face of jouissance and its structuring as the other side of the law. Kant proposes the establishment of a law which excludes any consideration of the relation between subject and object, and the capacity for the latter to produce pleasure or displeasure in the former, but rather is based upon the extent to which the subject's will is in accordance with an a priori law.

Following this logic, Lacan (2006a) is able to discern in the barbaric and licentious acts of Sade's libertines a certain adherence to a strict moral code which is articulated in the form of a maxim, which when enunciated takes as its foundation the acknowledgment of the other's supreme right to dominion over one's body, such that 'I have the right to enjoy your body", anyone can say to me "and I will exercise this right without any limit to the capriciousness of the exactions I may wish to satiate with your body' (p. 248). In highlighting the position of the enunciated "I" in this maxim as *not* the subject but the voice of law, Lacan proceeds to analyze its value as a universal and unconditional categorical imperative. Crucially it is the non-reciprocal nature of this edict that is significant. In the Sadean universe the right to jouissance is dependent upon the non-negotiable inequality between victim and aggressor in any sexual configuration and thereby all forms of social interaction. This however is complicated by the fact that the role of victim when occupied by those with a perverse structure is precisely to function as an object of the other's will in order to achieve jouissance, meaning that their pleasure is subjugated by their pursuit of the Other's enjoyment: to be an object of their drive (oral, anal, scopic, or invocatory).

As cultural fantasies about "sex after the Singularity" abound, we must ask what the future entails for these new configurations of sexuality and Artificial Intelligence? Whilst the likes of Elon Musk's Neuralink, Google's DeepMind and philosophers such as Nick Bostrom (2014) consider the implications of AI and Robotics for our legal system, culture, politics and human relationships, they fail to attend to the complex question of the *ethics of enjoyment*. Lacan's groundbreaking contribution to the ethical debate Kant avec Sade on the other hand, whilst well-used in the literature on psychoanalytic ethics, has yet to be employed in relation to the question of AI sex and its significance in human relationships.

Since the growing intervention of AI in social and sexual configurations dramatically changes the very stakes and scope of the law, it is a domain of ethics in complete overhaul. Furthermore, *Ghost in the Shell*, I argue, reveals the Sadean universe residing inside the seemingly most innocuous fantasies of Artificially Intelligent bodies that prevail in contemporary culture. As developers and research units seek to legislate for Artificial Intelligence and "Robot Ethics", the other side of the law as human jouissance, comes conspicuously into view. How, for example, do Asimov's famous three Laws of Robotics, which seem to resemble the current prevalent discourse on AI ethics, immediately provoke dissonance with Sade's ethical edict as mentioned above? We will recall that in *I, Robot* Asimov's (2018) Laws state:

1. A robot may not injure a human being or, through inaction, allow a human being to come to harm;
2. A robot must obey the orders given it by human beings except where such orders would conflict with the First Law; and
3. A robot must protect its own existence as long as such protection does not conflict with the First or Second Laws.

Lacan already demonstrated how the ethical law, when it comes to matters of human enjoyment, is very much more complicated that it may first appear, so how is this further problematized in the domain of AI? If

Sade's libertines' fantasies of a perpetual victim grow out of their fascination with the "second death", the inescapable law of castration, how would the "immortality" of the female Sexbot body, and potential for endless torment, feature as a mode of fantasy for the desires of the libertines? What kind of ethics can be built around the assumption of a subject who does not know castration and who, supposedly, can suffer indefinitely? Is the "Sexbot" the Sadean ethical imperative incarnate?

Lacan objected strongly to the idea forwarded by many thinkers after World War II (including the French publisher Jean Jacques Pauvert, Maurice Blanchot and Simone De Beauvoir among others) that Sade's libertine novels foreshadowed Freudian Psychoanalysis (Nobus 2019). Instead, as Dany Nobus points out, Lacan argued that Sade's works should be situated within the history of ethics and that he ought to be considered as moral philosopher above all else. He argues that:

> If there is a link at all between Sade and Freud it has nothing to do, then, with the former anticipating the latter but merely with the latter being able to formulate his fundamental 'scientific' concept of the 'pleasure principle' and especially its ostensible contradiction—the fact that one can experience pleasure in one's own and someone else's pain—because Sade had somehow prepared the ethical ground for it. (p. 115)

So, you may ask, what does this have to do with artificially intelligent female bodies (and the anxiety, fascination, and repulsion they provoke)? When taken to its speculative zenith, the *concept* of the Sexbot combines the *extimate* notions of enjoyment and the law via the challenge to subjectivity that Artificial Intelligence poses. The Sexbot furthermore presents us with the very kernel of the ethical foundation of the pleasure/pain dichotomy epitomized in the Sadean maxim which stipulates the other's right to enjoyment over ones' own body, and by extension our own compulsion to experience this submission to the other's will or our domination over it. For Sade, remember, to be in conformity with the moral law one must follow the maxim as outlined at the start, which contains within it an injunction to both victim and aggressor. As Lacan (2006a) posits in contrast to Kant's practical reason, the Sadean moral experience revolves entirely around jouissance, 'that by which Sadean experience is modified.

For it only proposes to instate itself at the inmost core of the subject whom it provokes beyond that by offending his sense of modesty' (p. 651). Lacan's point here is that as the object of the moral law is materialized in the figure of the libertine tormentor, it loses its Kantian inaccessibility (Nobus 2019). As distinct from the Kantian moral law as outside the realm of sensory experience, in the Sadean view the law is an abstract point of emission, which nevertheless presents itself as a disembodied voice, heard but not seen and always to be obeyed. Unlike for Kant, for the libertines however this disembodied voice of law is not God, as Kant could not possibly conceive of the possibility of the jouissance of God (ibid., p. 126). Since they themselves occupy the position of Gods, rather it is *nature itself* that determines their actions. As Dolmancé puts it, whilst waiting for his victim Eugénie to regain consciousness after a bout of torturous activities, 'if as merely the blind instruments of its inspirations, nature ordered us to set the universe ablaze, the sole crime would be to resist! And all the scoundrels on earth are purely the agents of nature's caprices' (Sade 2006, p. 168). The "tragedy" though for the libertines is that no matter how heinous or depraved their actions, their jouissance is but a pale imitation of the imagined enjoyment they would receive from executing the perfect crime; that is, of eternal suffering inflicted on their victims, along with their eternal ability to witness it and perhaps more fundamentally the fantasy of their own death. Of course, the obvious barrier to this possibility is the brute fact of the limitations of the human body and its ability to endure torment and destruction. So, as Lacan (2006a) puts it, the libertines have to admit 'the humility of an act in which he cannot help but become a being of flesh and to the very marrow, a slave to pleasure' (p. 652).

In other words, the libertines when all is said and done, can never achieve the full satisfaction they desire because it is always thwarted by the very human cycles of excitement and orgasm that are ultimately and inevitably always returning back to a state of equilibrium. So, could we not say that the ultimate pleasure for the libertine is in fact not just death, but immortality, *to be the undead*. The paradox for the libertines is that their bodily existence is both a source of unbounded enjoyment and also a barrier to the (fantasized) full and complete enjoyment of the 'beyond death'. As subjects of the symbolic or speaking bodies, the libertines can never really reach this state of plenitude and will always be subjected to the cycles of human pleasure, pain and eventually death. In which case perhaps the libertine would not wish to have a Sexbot, but *to be one*?

6.3 Undead Suffering

Ghost in the Shell adapted from the Japanese Manga series of the same name features the artificial intelligent female body and depicts its relationship to memory and trauma. The film allows us to examine the question of the body and suffering in relation to AI and ask how the Sadean imperative may help us to understand our fascination with the fantasy of an undead female body. *Ghost in the Shell* depicts a near dystopian future where virtuality and Artificial Intelligence have reached a state of sophistication such that the everyday texture of reality is interspersed with simulations and holograms, much like a walk through a Baudrillardian video game. The skyscrapers of what is in fact Hong Kong compete for dominance with giant holographic heads addressing the citizens with various commands, advertisements and provocations. Humans live alongside AI's in multiple forms of embodiment both humanoid and monstrous.

Like so many recent cinematic visions of AI, we are enthralled by a beautiful feminine protagonist; in this case, it is Scarlett Johansson portraying the character of anti-terrorism operative *Major Killian*. Killian is supposedly neither human nor AI. After an accident which destroyed her human body, she has been reanimated from the merging of her brain with an entirely synthetic body. She is presented to us as a perfect specimen: eternally young, beautiful, strong, perpetually naked and, of course, alabaster white.[2] According to the CEO of *Hanka Robotics*, Killian is a weapon in the fight against the threat of a new kind of cyber-terrorism which can hack into AI and human brains and perform mind control. She is told that due to her unique combination of human and

[2] Given the original Manga story was set in Japan the character of Major Killian was of course Japanese, leading to accusations of Hollywood whitewashing in the casting of Scarlett Johansen. The defence was given that, since the body of Killian was augmented, she was therefore not bound to being Japanese. Which ultimately only served to highlight the implicit superiority given to the white body as chosen for Killian's reincarnation.

non-human qualities she represents a new dawn for civilization. In effect Killian is a last bastion against the complete algorithmic takeover of humanity, whose relevance and efficiency is waning.

Given the progressive redundancy of the fragile and fallible biological body, Killian is, her designer Dr. Ouelet tells her, 'what we will all become'. Noticing Killian is experiencing distress, her doctor is concerned for Killian's psychological wellbeing, even though she suffers multiple physical assaults which leave her temporarily incapacitated she does not seem to feel any bodily pain. Suffering from occasional glitches in her own memories, she starts to experience visions of what she is led to believe are faults in her programming. When she first awakes from her transformation into a cyborg (or Sexbot), Killian asks why she can't feel her body. Dr. Ouelet explains that her body could not be saved after a tragic boat accident which killed her whole family, and that she now has an entirely synthetic, yet supercharged new shell. Her brain, however, is completely intact.

Major Kilian is sent to hunt for a hacker who is terrorizing Hanka Robotics. After a robotic geisha is hacked and starts a killing spree at a Hanka business conference Killian is sent in to "neutralize" the Geisha. After which she defies protocol and decides to take a dangerous virtual deep dive into the AI of the defunct Geisha to see what she can retrieve from its memories. Here she discovers the author of the hack: Kuze. After Major Killian is eventually captured by Kuze, he reveals that he was himself a test case of the same type as her, and there were many other before her. She discovers that the story she was told about her "origins"—i.e. that her life was saved after an accident and her brain uploaded into a new, more sophisticated, non-biological body—was all a lie. In fact, Killian's life was "stolen". It becomes clear though, through the visions she experiences as "glitches", that her own memories have resisted complete annihilation after her organic brain was uploaded in her new synthetic body. It turns out she and Kuze were in fact young anti-augmentation activists who had run away from home. Posing a threat to political order, they were killed by a new and menacing technocratic regime that erased their memories in the hope of turning them into ultimate fighting machines in the service of the state. In revenge Kuze wants to create a super network of human-AI consciousness all connected to a central "brain" and implores Killian to join with him. She refuses, however, intent on retrieving her subjectivity and goes about a mission to recover her lost memories. Major Killian holds out the possibility of a refusal of authority over her remaining humanity, albeit bolstered by a superhuman cybernetic body. Through it all, Killian attempts to retains hold of her subjectivity and resists total assimilation into an automated and machinic life form. But what may we glean from the depiction of the augmented female body in its relation to questions of the subjectivity of Killian? How does the film deal with the problem of Killian's "lost" past? And what are we to make of the fact that whenever she engages in combat, she mysteriously loses her clothing and fights with a completely bare silicone body? Is it for her own pleasure or for her opponent's? What is Killian's relationship to her body? Does she feel physical sensation? And if not, how does she operate in a physical world if her body can feel nothing? In other words, in what ways does the character of Major Killian speak to the question of sexuation in relation to the Sadean Universe of *undead* enjoyment?

It seems that what is retained by Killian after her reanimation is her *subject position*, an indelible stain in the fabric of reality that cannot be substituted nor lost no matter what memories (conscious or unconscious) are erased by her physical designers. However, she appears strangely devoid of enjoyment, given that she cannot feel any physical sensation. But is this really the case? Can it be that Killian does in fact enjoy? And if so, in what is this enjoyment constituted? Like so many depictions of female AIs, is what we are in fact being asked to imagine a fantasy of *feminine jouissance*? An unbounded jouissance that defies symbolization?

This is just one of the many cinematic instances where the female body is put to work in pursuit of an answer to the question of the relationship between sex and the law. Major Killian's subjectivity is of course intimately bound to her embodiment. Hence why, only when she is fulfilling her purpose as "ultimate weapon", she is naked and eroticized? A trite point about female representation in cinema perhaps, but there is something more significant behind this. Major Killian is not just a sexualized female body, she is a super-human one, and in examining the idea of the superhuman, we may thereby understand something about the Sadean ethics of sexuality and its relationship to birth and death. Ultimately

Killian's primary source of suffering still revolves around an originary loss. An impossible object that her reincarnated body mourns after. It is no surprise then that the film ends with Killian being reunited with her mother, her original lost object.

6.4 The Lamella: Lost Enjoyment

This fantasy of the 'undead' lifeform that *Ghost in the Shell* depicts is described by Lacan in *Position of the Unconscious* (2006b) and *Seminar XI* (1977) where he speaks of the mythical *l'hommelette* or "manlet". He then further characterises it as the *lamella*, that strange amoeba that leaves the body at the time of birth when the child is separated from the placenta. He asks us to imagine a phantom 'infinitely more primal form of life' that would take flight away from the new-born (p. 717). This crêpe-like form is the remainder of the subject before it becomes sexed:

> Whenever the membranes of the egg in which the foetus emerges on its way to becoming a new-born are broken, imagine for a moment that something flies off, and that one can do it with an egg as easily as with a man, namely the hommelette, or the lamella. The lamella is something extra-flat, which moves like the amoeba. It is just a little more complicated. But it goes everywhere. And as it is something—I will tell you shortly why—that is related to what the sexed being loses in sexuality, it is, like the amoeba in relation to sexed beings, immortal—because it survives any division, and scissiparous intervention. And it can turn around. Well! This is not very reassuring. But suppose it comes and envelopes your face while you are quietly asleep...
>
> It is the libido, qua pure life instinct, that is to say, immortal life, irrepressible life, life that has need of no organ, simplified, indestructible life. It is precisely what is subtracted from the living being by virtue of the fact that it is subject to the cycle of sexed reproduction. And it is of this that all the forms of the objet a that can be enumerated are the representatives, the equivalents. (Lacan 1977, pp. 197–198)

So, the lamella has no sensory system, in other words it has no need for partial drives oral, anal, scopic or invocatory, synthesizing all these aspects into one complete plenitude of pure satisfaction, wholeness and presence. It exists purely in the real with no need for symbolic mediation, and 'thus has an advantage over us men who must provide ourselves with a homunculus in our heads in order to turn that real into a reality' (Lacan 2006b, p. 717).

The manlet or lamella is indestructible, immortal and undead. In other words, the lamella is libido. Pure enjoyment; a logical impossibility of course, yet whose originary loss provides the formal conditions for the structure of the sexed being. The lamella is the undead life force that the prosthetic god attempts to replicate but will never capture. The myth of the lamella, we could say, is the very thing that inhabits the eschatological fantasies of the Singularity, that moment when humankind is replaced by an immortal and indestructible digital form of life. This problematic phenomenon is arguably what we find manifested in the *lathouse*, that which, paraphrasing Lacan, is *not quite being and not quite the other* (Lacan 2007).

The relationship between technological forms of life and sexuality is indexed then, upon a certain mode of enjoyment or regime of governace. As *Ghost in the Shell* epitomises, the character of Major Killian serves as a perfect vessel to reunite the feminine subject back with the lamella, in the guise of an undead silicone fighting machine. But ultimately this, as with all attempts at breaching castration, fails. So, what is it in the subject that remains indestructible? Is it the lamella? If Major Killian retains her subjectivity despite the complete replacement of her body and her memories are virtually replaced, in what sense is she traumatised by the event of her physical death?

Here it is useful to bring in Malabou's and Žižek's debate over the post-traumatic subject. In *The New Wounded* Malabou (2012) criticizes the Freudo-Lacanian paradigm of unconscious trauma on the basis that it cannot possibly grasp the radical change of a subject who has faced a massive brain injury which effectively erases all memory and as it were "resets the program". In this case, she argues, it would be impossible to apply the logic of Freudian trauma which operates via a double inscription. That is to say, the initial occurrence of the trauma is not registered as trauma for the subject but only becomes traumatic when a subsequent experience imbues this previous event with meaning and causes suffering to the subject. The error she is making according to Žižek is that, in focusing so much on the traumatic content of the supposed erasure of all memories, she omits the trauma of the erasure of all positive content that *is subjectivity itself*. In other words, a radical trauma in the form of a massive brain damage would reveal the pure empty form of subjectivity. The form which remains when all positive content is removed:

> precisely insofar as it erases the entire substantial content, the traumatic shock *repeats* the past, i.e. the past traumatic loss of substance which is constitutive of the very dimension of subjectivity. *What is repeated here is not some ancient content, but the very gesture of erasing all substantial content.* This is why, when one submits a human subject to a traumatic intrusion, the outcome is the empty form of the 'living-dead' subject… [W]hat remains after the violent traumatic intrusion into a human subject which erases all its substantial content is the pure form of subjectivity, the form which must have already been there. (Žižek 2016, p. 339)

So, when Major Killian awakes from her traumatic experience of total brain erasure and complete bodily substitution, the trauma she is exposed to is not the loss of subjectivity but in effect the stripping out of her objective substantialized content revealing the empty form of her as subject. When the memories of her past life appear to her they are traumatic insofar as they intrude as if from nowhere into the empty space of subjectivity. Killian is suddenly exposed to the effects of castration, moving from undead subject back to the trauma of birth and the realm of the living.

To return to Sade's (2006) *Philosophy in the Boudoir*, the action also centers around the complete erasure of a previous form of subjectivity and the explosive discovery of new forms of jouissance of the "exquisite" female protagonist Eugénie de Mistival. It is significant furthermore that her primary cause of suffering and indeed the victim of the culmination of her most depraved fantasy is her own mother. It is her mother whose unbearable (and hypocritical) virtuousness causes Eugénie to be caught between her own so-called "natural" desires and passions and the restrictions put upon her by polite society. As Dolmancé explains to her during her sexual "education":

> Did her mother think about Eugénie when she brought her into the world? The hussy let herself get fucked because she enjoyed it, but she was quite far from envisioning a daughter. So let Eugénie do whatever she likes to that woman! Let's give her free rein, and let's content ourselves with assuring that no matter how extreme her excesses, she'll never be guilty of a crime. (p. 57)

Horrifyingly it is with the rape and torture of her mother that Eugénie supposedly fulfills her ultimate desire. Whilst of course Killian does nothing of the sort, the character of Dr. Ouelet who fulfils the role of Killian's new mother being her "designer", is blamed and killed by Hanka CEO once Killian has "gone rogue". But Killian's relationship to her 'real' mother is one of pure enigma. Her biological mother ties her to her human mortality and her indelible subject position as 'stain on reality', yet her second 'prosthetic' mother Dr. Ouelet, redesigns her and facilitates her escape from the second death of the law of castration (much like Madame de Saint-Ange attempts to "redesign" Eugénie in line with different laws). Her new body allows Killian to live outside of the restrictions of pleasure and pain which her biological body as given to her by her first mother could not accommodate.

As revolting and brutal as the appetites of Sade's libertines are, we may nonetheless see some of their ontology present in film *Ghost in the Shell.* What kind of fantasy victim would Kilian represent? A body that can't die yet can suffer indefinitely generating an unquenchable jouissance unfettered by the limits of human biological cycles? And what kind of subject is Killian? Does she have a history? And does this form the basis of her suffering and her enjoyment? Is this not the Sadean ethical dream of ultimate satisfaction? Is this the pervert's dream? Or the Sadist's dream? As pointed out by Nobus (2019) at no point does Lacan in fact directly equate the Sadean ethics with perversion which is mentioned only once in the text in a superficial manner and Sadism while mentioned several times is not treated as homologous to a Sadean ethics even if the brutal psychoanalytic category bears his name, these are assumptions taken up later by other theorists and analysts.[3] Lacan's aim in Kant avec Sade is to

[3] Notably Miller (1998) and Žižek (2016).

complexify the tripartite relation between the subject's enjoyment and suffering on the one hand and knowledge of the other's pain and pleasure on the other and how *the law* mediates between the two.

In Sade's dark satire on contemporary French moralizing and its class politics, Eugénie ends the story in spectacularly gruesome fashion by sewing up the vagina and anus of her mother after she is raped by the syphilitic gardener. It is undoubtedly infinitely more graphically and explicitly violent towards the mother than *Ghost in the Shell*, but what seems to be at stake and under erasure in both stories, is the position of the mother as sole progenitor and indeed moral guardian of the species. For Sade the mother as the holy grail of religious discourse and morality must be desecrated, whilst for *Ghost in the Shell* the mother occupies an ambivalent role, supposedly the origin of Killian's subjectivity but ultimately limiting to the progression of the "species". The question of the mother and *reproduction* is an area highly under theorized in Artificial Intelligence debates, one which I will explore further in the next chapter.

This sketches out for us another crucial factor in the Sadean ethics which hinges on male and female sexuation. The victim, for the libertine must be female, but why? Because the female subject represents the ultimate empty subject, for whom substantialised content is ontologised by the carnality of the female form. It is precisely the hyper-materiality of the female body that acts as veil for the negativity of being which the Sadean libertine cannot bear. As Žižek (2016) puts it:

> This redoubling of the body into the common mortal body and the ethereal undead body brings us to the crux of the matter: the distinction between the two deaths, the biological death of the common mortal body and the death of the other "undead" body: it is clear that what Sade aims at in his notion of a radical Crime is the murder of this second body. (p. 334)

What Sade missed and Lacan realised, Žižek argues, is precisely that these two deaths come in reverse order: 'I can see that the second death comes prior to the first and not after as de Sade dreams it' (Lacan cited by Žižek, p. 335). For Sade's libertines (not Sade himself, as Žižek will hypostatize him) the universe is pure substance without subject, they still believe in the big Other and 'Nature as ontologically consistent realm' (p. 335). Therefore, according to Žižek:

> Sade continues to grasp reality only as substance and not also as subject, where subject does not stand for another ontological level different from substance but for the immanent incompleteness—inconsistency—antagonism of Substance itself. (p. 335)

If we see Killian as our ultimate fantasy of the undead female body, it seems that she embodies the irreconcilable trauma of subjectivity that artificial life would 'suffer' from. Killian is both the indestructible killer and the perpetually killed. The law and enjoyment as first problematized by Lacan's reading of it, is brought to a strange conclusion in the figure of the undead body that seems ubiquitous in both our sci-fi fantasy worlds, and inevitably soon our legislation on embodied Artificial Intelligence. A patipolitical regime of governance perhaps? Major Killian may be augmented into the form of a quasi-invincible non-biological body, but yet her "humanity" appears precisely at the point where satisfaction fails. In her search for the lost memories that escape her grasp, the voices she hears, the images she sees in her technological "glitches" point to a structure of fantasy that yearns after various lost objects, or one in particular. The prosthetic god that Freud once postulated was one which also dreams hubristically of not suffering the effects of castration, does not die and is not born. From the point of view of the Sadean Universe, Killian is probably the ultimate victim, a futuristic Eugénie de Mistival. Not only is she an impeccable body of alabaster virtue, perpetually unscathed and virginal, yet inhumanly strong but she also has the capacity to suffer indefinitely and probably can't die. Is this what we could call the start of a Sexbot ethics?

Bibliography

Asimov, I. (2018) *The Complete Robot*. London: Harper Collins.
Baudrillard, J. (2008) *The Perfect Crime*. London: Verso.

Bostrom, N. (2014) *Superintelligence: Paths, Dangers, Strategies*. Oxford: Oxford University Press.

Chiesa, L. (2007) *Subjectivity and Otherness: A Philosophical Reading of Lacan*. Cambridge, MA: The MIT Press.

De Sade, D.A.F. (2006) *Philosophy in the Boudoir*. London: Penguin.

Kant, I. (2002) *Critique of Practical Reason*. Indianapolis: Hackett Publishing Company.

Lacan, J. (1977) *The Seminar of Jacques Lacan Book XI: The Four Fundamental Concepts of Psychoanalysis*. London: Karnac Books.

Lacan, J. (1993) *The Seminar of Jacques Lacan Book III: The Psychosis 1955–1956*. London: Routledge.

Lacan, J. (1998) *The Seminar of Jacques Lacan Book XX: Encore—On Feminine Sexuality, the Limits of Love and Knowledge 1972–1973*. London: W.W. Norton & Company.

Lacan, J. (2006a) 'Kant avec Sade' in *Écrits*, pp. 645–670. London: W.W. Norton & Company.

Lacan, J. (2006b) 'Position of the Unconscious' in *Écrits*, pp. 703–721. London: W.W. Norton & Company.

Lacan, J. (2007) *The Seminar of Jacques Lacan Book XVII: The Other Side of Psychoanalysis*. London: W.W. Norton & Company.

Lacan, J. (2019) *The Seminar of Jacques Lacan Book VI: Desire and Its Interpretation*. Cambridge: Polity Press.

Land, N. (2011) *Fanged Noumena: Collected writings 1987–2007*. New York: Urbanomic.

Lotringer, S. (1988) *Overexposed: Perverting Perversions*. Los Angeles: Semiotext(e).

Malabou, C. (2012) *The New Wounded: From Neurosis to Brain Damage*. New York: Fordham University Press.

Mbembe, A. (2003) Necropolitics. *Public Culture* 15(1): pp. 11–40.

Miller, J-A. (1998) Sobre "Kant con Sade". In *Elucidation de Lacan: Charles Brasileňas*. Buenos Aires: Paidos.

Nobus, D. (2019) 'Kant with Sade' in S. Vanheule, D. Hook & C. Neill (Eds.), *Reading Lacan's Écrits: From 'Signification of the Phallus' to 'Metaphor of the Subject'*, pp. 110–167. London: Routledge.

Richardson, K. (2018) Campaign Against Sex Robots. Available (01.03.20) on: https://campaignagainstsexrobots.org/.

Žižek, S. (2016) *Disparities*. London: Bloomsbury.

Zupančič, A. (2000) *Ethics of the Real: Kant, Lacan*. London: Verso.

Filmography

Kleeman, J. (Creator) (2017) Rise of the Sex Robots [Documentary]. UK: The Guardian.

Sanders, R. (Director) (2017) Ghost in the Shell [Film]. US: Paramount Pictures.

7

What Can I Hope For? Reproduction, Replication, Immortality

What becomes of a master without a slave? He ends up terrorizing himself. And of a slave without a master? He ends up exploiting himself. The two are conjoined today in the modern form of voluntary servitude: enslavement to data systems and calculation systems- total efficiency total performance. We have become masters—at least virtual masters—of this world, but the object of that mastery, the finality of that mastery have disappeared.
—Baudrillard (2008, pp. 113–114)

7.1 Baby X

The third Kantian question 'What may I hope for?' emerges first in his second critique but develops fully in his (1987) third *Critique* in which Kant not only undertakes an analysis of the sublime which grows out of his work in the previous two critiques but asks after the purpose of man. By the end of the first critique Kant had established certain fundamental antinomies of human reason. Most significantly that of causal determinism, which allows empirical science to work towards a fully synthesized account of the cause of all events in the world and on the other hand, that of spontaneous causality which is the domain of freedom and ethics. The

I. Millar, *The Psychoanalysis of Artificial Intelligence*, The Palgrave Lacan Series,
https://doi.org/10.1007/978-3-030-67981-1_7

third critique therefore constitutes a discussion of the overlapping of these two domains in the faculty of judgment. Both the Understanding (*Verstand*) and Reason (*Vernunft*) must be invoked in this realm. The third critique is concerned with teleological judgement, and the concepts of biology and heredity, in which Kant will argue that man is the ultimate telos of nature due to his capacity for reason. In another work *Anthropology History and Education* (2007) Kant strongly objected to the proto-eugenicist idea forwarded by Pierre Louis Moreau de Maupertius that in the service of the perfectibility of human nature we should be able to determine in advance a beings physical or cognitive ability. Kant instead argued for the inherently ethical dimension to how science and by extension the moral subject must deal with the contingency of nature and its good and bad elements. For Kant then the question what may I hope for was inextricably tied to our faith in the possibility of free will, the immortality of the soul and ultimately the existence of a non-deceitful God who would supposedly have designed the world according to principles of justice. Whilst for Kant we cannot have knowledge of these things, consideration of the moral law necessarily results in a justification of our belief in them. Ultimately for Kant ethics inevitably leads to religion which culminated in the positing of God.

Naturally however I approach the third Kantian question not in the manner of Kant but in the manner of Lacan. Following on from the previous chapter, once we have reconceived of the moral law according to Lacan's reading of Kant with Sade, the possibility of the immortality or indelibility of the subject, the question of justice (as a function of jouissance) and the will—conceived of in Sadean terms as following desire to the end—takes on an entirely new dimension, one that leads us to consider the notion of the undead body and its relationship to enjoyment. Furthermore, the idea of God as guarantor of the immortality of the 'soul', as Lacan elaborates in *Seminar XX* is conceptually and logically bound to woman, the phallus and sexuation. And nature seen from this perspective takes on a completely different importance. As this chapter will argue, the question of the destiny of the human species is significantly complicated by the possibility of the passage from reproduction to replication.

This Kantian question viewed from the Lacanian position prompts us to reevaluate the question of sexuation in relation to reproduction as a form of human immortality and the problem of sex for the human subject. We may ask, what is the function of the semblant of the telos of life as reproduction? And if the semblant is removed what becomes of the traditional structural questions of filiation in terms of the role of mother and father in the psyche? Freud already wondered how society would change when humans realised that they don't have to intend to procreate (even unconsciously) to have sex, but surely the new theoretical question for psychoanalysis is; what happens when humans *don't ever need to have sex at all*, to procreate? Admittedly this is already the case with in-vitro fertilization but more radically, what is the status of the reproductive body when we can gestate embryos via *ectogenesis*; outside of the womb? Indeed, the technology is already more evolved than our current medical or indeed psychoanalytic ethics have begun to consider. The development of the artificial womb has already begun, which will surely have profound implications, for women's reproductive rights, and the rights or legal status of unborn fetuses, but also for sexuation more generally. The enigma of the reproducing female body is arguably one of the fundamental impasses underlying the whole of the psychoanalytic enterprise. Freud's two existential questions said to be characteristic of either side of the neurotic dialectic, being on the one hand; 'am I dead or alive?' and on the other; 'am I a man or a woman?' Both questions stem from the impossibility of explaining firstly the mystery of life as creation *ex nihilo*, and secondly; the curious "magic" that grants some bodies the power to create another subjectivity out of nowhere. In *Sexuality in the Aetiology of the Neurosis* Freud (1898) contemplates the possibility of a future where sex and procreation were not necessarily bound and that the risk of (or desire for) pregnancy was not always looming in the background of any sexual encounter. For Freud this development would surely bring about fundamental shifts in our social conditions, not to mention for women's sexual freedom. We may look back at Freud as naïve in this respect, but it would still be fifty years before the first contraceptive pill was invented in 1951. Then in 1956–1957 *Seminar IV* Object Relations, Lacan, surprised by the news that a woman has used a frozen stock of sperm from her deceased husband in order to inseminate herself, asks the question; *What is a father?* Concluding that it is precisely the dead father, which is the symbolic father, or in Lacanian terms the Name-of-the-Father, a notion nonetheless already present in Freud (as totemic father of the horde). Much later in *Seminar XVII* (perhaps unconsciously referencing the reserve sperm of the dead father that he found so intriguing some 12 years earlier) in direct reference to *Totem and Taboo*, Lacan (2007) states: 'equivalence is therefore drawn, in Freudian terms, between the dead father and jouissance. It is he who keeps it in reserve, if I can put it like that' (p. 123).

He continues:

> Here the myth transcends itself through stating in the name of the real- for this is what Freud insists upon, that it actually happened, that it is the real—that the dead father is what guards *jouissance*, is where the prohibition of *jouissance* started, where it stemmed from. (p. 123)

With the dismantling of the previous unconscious structures of filiation that contemporary techno-science has now begun to offer, in the form of *in-vitro* fertilization, stem cell research leading to the possibility of genetic material from more than two parents, and various forms of surrogacy, not only can a symbolic father be 'dead' but now so can a mother. In other words, the mother transcends her previous role as bound to the material, viceral and emotional labour of maternity. Regardless of the manner in which a child is born (or made) from a living or dead father, conceived inside or outside a womb, by a surrogate, from a donor egg, the question around which the problem of filiation revolves in psychoanalytic terms is that of the *primal scene*. In other words, for the psychoanalytic subject the deadlock which is exposed in this problematic is the moment of their own creation, a moment which structurally they cannot know anything about. The most radical form of this deadlock psychoanalytically comes to us in the form of the possibility of asexual reproduction, that is to say *replication*.

The epigraph of this chapter appears in an essay entitled *The World without Women* which Baudrillard borrows from the title of a book by Virgilio Martini *(Il Mondo senza Donne 1935)* in which the world is decimated by a mysterious illness which has been designed to wipe out all childbearing women between puberty and menopause. As Baudrillard points out, the book was written some 50 years before the outbreak of AIDS but provides some uncanny resonances with the fears and fantasies surrounding the real life unfolding of the AIDS epidemic in the 1980s; in the story the disease broke out in Haiti and is purportedly a plot hatched by homosexuals to exterminate women. Essentially the book as Baudrillard sees it is an allegory of the extermination of all otherness, for which the feminine is the metaphor. In his view, the deeper allegory is not in fact that of AIDS but more fundamentally the virus that all humans have fallen victim to, that is the 'virus destructive of otherness' (Baudrillard 2008, p. 111). Though for the moment this virus 'does not affect the biological reproduction of the species' what it aims at is 'the symbolic reproduction of the other, in favour of 'cloned asexual reproduction' (ibid.). Later in *The Vital Illusion* Baudrillard argues that the advent of human cloning, far from signaling a new age of immortal life for the human species, paradoxically brings about the end to humanity as we know it. The drive towards the Same he fears, is killing us.

But what does cloning or replication mean for humans? Is it the end of reproduction? Is it the end of sex? Is it the end of the child? Dr. Mark Sagar, CEO of Soul Machines, a New Zealand-based company that develops intelligent, emotionally responsive avatars has invented Baby X, a virtual nervous system with the avatar of a human baby which is being trained to respond to the world "like a real human". The uncanny feeling of wonder at seeing the AI child's face respond to learning new words and identifying images brings to mind once again the vulnerability and supposed humanity of 2001s Hall 9000. But even though we know the baby doesn't really *feel* anything, there is something particularly unsettling about the use of the child in the context of machine learning. What is it precisely? In 'Note on the Child' Lacan (2018) writes:

> If the gap between the identification with the ego ideal and the piece taken from the mother's desire lacks the mediation that is normally provided by the father's function, it leaves the child susceptible to every kind of fantasmatic capture. He becomes the mother's "object" and his sole function is to reveal the truth of this object. The child realizes the presence of what Jacques Lacan designates as objet a in fantasy. By substituting himself for

> this object, the child saturates the mode of lack whereby (the mother's) desire is particularized, whatever her specific structure—neurotic, perverse or psychotic. (pp. 13–14)

For Lacan as we know, the child has a pivotal role for the mother; realizing the object a in fantasy. The child therefore as symptom of the family structure is in a sense the only real incarnation of *object a* that can exist. But the child has this function not just for the mother but arguably for the social bond in general. The child is in theory the last sacred object that unifies all cultures, the only human that legally and morally must always be protected above all else. But precisely because of this, the opposite often happens. The notion of 'the child' is universally cherished even when it is most defiled and abused in reality. Perhaps this is why the generation of an AI child *Baby X* at Soul Machines is so unnerving and strange. It incarnates this paradox between innocence and omnipotence that the figure of the child already holds within it. A child simultaneously 'knows nothing' and must be taught how to *be in the world* and also is the bearer of a potentially unlimited capacity to learn, think and create.

So, what form does the figure of the Sexbot take in relation to these potential new conditions of filiation and in answer to the question *what can we hope for*? In Chap. 5 the Sexbot as represented by Ava in the film *Ex Machina* was an entirely constructed body and brain, one whose existence was predicated on semblance of a biological body; an exterior iteration of AI. In Chap. 6 the Sexbot depicted by Killian in *Ghost in the Shell* is one who was born in a human body and then transferred to a synthetic body; an intimate iteration of AI. In *Blade Runner 2049* however, whilst the obvious figure of the Sexbot may appear to be the character of Joi the holographic girlfriend, it is in fact Blade Runner K. K is the final *aufhebung* resulting from the transition from built Artificial Intelligence to reproducing Artificial Intelligence, to born Artificial Intelligence; an extimate iteration of AI. K, as the possible child of a replicant presents us with the question of non-biological reproduction and filiation.

7.2 Blade Runner 2049: What Is a Father?[1]

Much like the first *Blade Runner* movie, Denis Villeneuve's (2017) *Blade Runner 2049* sequel seeks to ask fundamental questions about the nature of human existence in the context of accelerated techno-capitalism, unstoppable space colonization, and growing anxieties about the technological domination of all human life. Whilst the first film provoked us to doubt the reliability of our own experiences in a sort of post-modern Cartesian or *Deckardian* meditation, so the sequel also entertains the theme of radical Cartesian doubt in its posing of the question of K's implanted memories. This time though, the added layer of psychoanalytic significance that *Blade Runner 2049* presents us with, is the problem of our protagonist's progenitor. After the discovery of a missing child, the product of Deckard and Rachel's union in the previous film, the young blade runner *K* is tasked with tracking the child down and retiring it. K however, starts to think that the missing child might be him. He may not be the creation of Wallace Corporation as he had always believed, opening up the possibility that he was the product of sexual reproduction; that he had a childhood and a family. It would seem the crux of the protagonist's dilemma is no longer "am I human?", but "was I born?" *Blade Runner 2049* sees the transition from the traditional epistemological question tackled by its predecessor, the original *Blade Runner* to what is arguably a slightly different, perhaps even more strictly psychoanalytic problem relating to Artificial Intelligence; that is the question of birth, knowledge and castration. And significantly the figure of the child.

In this rendering of the tale, K, in searching for the film's object of desire—the missing child—fantasmatically tries to reconstruct his very own primal scene. Whilst in the previous film the replicants didn't know they weren't human and struggled with the realization that they were not "real", K and his generation of Nexus 9 models by contrast, know now only too well the limited nature of their existence. For this reason, when K discovers he may actually have been born and is thus "special" he starts

[1] Elements of this section appear in my forthcoming chapter 'Before we even know what we are we fear to lose it: The Missing Object of the Primal Scene' in Calum Neill's (ed.) (2021) Lacanian Perspectives on *Blade Runner 2049*. London: Palgrave Macmillan.

to believe he is the "chosen one", born of the union of man and woman. The enigmatic and profoundly human problem of sexuation is thus brought to the fore.

The anxiety depicted here is about the disintegration of the sexual relation. Since in this imagined future, humans are reproduced "artificially" and without the need for the work of female gestation, the significance of sexuation becomes ever more conspicuous. Why would we even still need so called "men" and "women" if we can reproduce a-sexually we may ask? Lacan (1992) states in *Seminar VII*:

> The idea of creation is cosubstantial with your thought. You cannot think, no one can think, except in creationist terms. What you take to be the most familiar of your thought, namely, evolutionism, is with you, as with all your contemporaries, a form of defence, of clinging to religious ideals, which prevents you from seeing what is happening in the world around you. But it is not because you, like everyone else, whether you know it or not, are caught up in the notion of creation, that the creator is in a clear position for you. (p. 156)

Lacan will go on to connect the question of creation with the Freudian dead father and more specifically the drive. Which he calls 'an absolutely fundamental ontological notion, which is a response to a crisis of consciousness that we are not necessarily obliged to identify, since we are living it' (p. 157). It is thus the idea of creation *ex-nihilo* that plagues the subject, and that always leads him back to look for his own cause.

Whilst the first film was concerned with the epistemological question of the analogue and the digital and the transition between the two as marking a loss of reality, epitomised by the Cartesian "what is real?"[2] The second film is more concerned with the onto-thoelogical question "what am I?".

---

[2] As Flisfeder (2017) has commented of the original Blade Runner and its fascination with playing with notions of authenticity and reality, not only does the film deal explicitly with the concept of simulacrum, but the film *itself* as a cultural object, owing to its multiple reworkings (7 in total) is a simulacrum. Flisfeder remarks: 'In this sense Blade Runner is always a simulacrum of itself with each version marking and adding a new layer to the historicity of its form, the original version seems to matter less and less' (p. 97).

Lacan's rethinking of Freud's understanding of biology was crucial to his subversion of psychoanalytic practice and theory. By reusing Freud's radical ideas to reconceive of the significance of biology for psychoanalysis, Lacan took Freudian biology (which was working with the scientific paradigms of the day) and 'rebooted' it. As he noted 'Freudian biology has nothing to do with biology' (Lacan 1988, p. 75).

In this way we may understand *Blade Runner 2049s* interest in the question of replicant biology as a return to the Freudian concern with myth and impasse that first inspired Lacan's structuralist reading of Freud's ideas. K's desire for knowledge of his memories and birth is indicative not just of a form of radical doubt over the nature of reality and the constitution of human subjectivity—which during the heyday of postmodernism was fuelled by a Baudrillardian speculation about the simulated constitution of our everyday lives (as we see in the first film)—but furthermore, K's predicament is articulating the structure of the primordial loss constitutive of the psychoanalytic subject. One that is inherently bound to birth, language, knowledge, enjoyment and the body.

The object cause of desire in this film then is of course the child, both for the sake of the warring factions (the replicant revolutionary front and Wallace Corporation) who wish to either protect it or destroy it and capitalise on it, but on a more psychoanalytic level, the child is K's object cause of desire in that he wants to be it, it is his *manque à être*. K seeks the ultimate knowledge of the primal scene with his burning question; "where did I come from?" In the quest to find the lost child of Rachael, K is unwittingly "searching for himself" like all tragic heroes. As distinct from the previous generations of replicants, who were at pains to discover that their memories were not their own, K has the opposite problem. Of course, once he gets the idea that his memories may not have been fabricated, but in fact are "real", he retroactively starts to imagine a loss of something that he never knew he had, that is, a mother. This primordial loss is of course the quintessential mark of the human subject, a retroactive imagining of an impossible plenitude through absolute satisfaction.

What we have here then is a strange sort of "interpretation in reverse", as Miller (2007) would call it. Instead of starting off with the symptom and working backwards to a trauma, we are starting off with the insertion of a trauma in order to give K a symptom, that would in a sense *make him human*. As opposed to Freud's first clinical invocation of the trauma of the primal scene in his treatment of Sergei Pankajeff—a.k.a. the Wolfman (Freud 1918)—K is devoid of a neurosis that would point to any form of fantasy of origins. So, the postulation of the missing child retroactively creates a historical connection to his birth and an imagined relationship between father and mother that pre-existed him; a 'desire of the other' that frames his own feelings of exclusion.

Is this what K needs for him to enter into the realms, not just of real humanity, but of real 'masculinity'? Of course, who does he come across in his quest for the child but the exiled *pater familias*, grisly old Deckard. And naturally they must beat each other within an inch of their lives, since as we know you can't be a real (hu)man unless you have at least *tried* to kill your father, unconsciously or otherwise. So, K's life is a simulacrum of humanity and of masculinity. His girlfriend Joi is a hologram who supposedly gives him everything he wants at the touch of a virtual button. In Lacanian terms she is literally the simulacrum of his symptom. If woman is man's symptom, then in this case Joi the hologram is by extension the symptom of K the replicant. Joi performs the ultimate male fantasy of a woman who exists only for the pleasure of her man and can transform herself, her mood and her outfit to suit his desires.

K's boss Lt. Joshi on the other hand is the mother figure—and representative of the conservative values of the traditional family, intent on creating boundaries and borders—who it seems, structurally speaking, is sending him to his own annihilation. In a sort of twisted Oedipal logic, K seeks the object of his primal scene and finds out he is far too close for comfort. As K becomes aware of the ramifications of his situation, he has to come to terms with the fact that at the very least he has a sister, in the form of Dr Anna Stelline. One of them is the copy of the other, however, and K assumes, given the poignancy of his wooden horse memory, that it is she who is the replicant of him. Unfortunately for him, he is mistaken.

Of course, it is not for nothing that *Joi*, the embodiment of feminine perfection, is K's companion throughout the film. And, in fact, it is Joi who convinces him of his exceptional quality. She encourages K to believe that he is different from the rest of the replicants. She tells him he must be 'a child of woman, born, wanted, loved', and that, furthermore, he needs a name: 'Joe'. Is this a Kafkaesque Joseph K. or a biblical Joseph?

The former points to K's entrapment in the alienation of bureaucracy and the latter to his potential importance in the genesis of a replicant people, as we shall see.

But why is Joi so sure about this? She, as his algorithmically programmed fantasy, is designed to tell him what he wants to hear. In other words, her existence props up his masculinity. In formal terms, masculinity literally conforms to the belief in the exceptional quality of one's being. To be a man one must belong to the category of "men", which entails the phantasmatic positing of an ultimate "Man". Recall in the final scenes of the first Bladerunner film, as *Deckard* has finally vanquished the last remaining rogue replicant *Roy Batty*, his colleague *Gaff* tells him 'you have done a man's job'. The ambiguity here, alludes not only to the ongoing mystery about Deckard's status as human or replicant, but also more fundamentally belies the structure of masculinity itself; partaking of that elusive and superhuman quality of "manliness". Joi's position conversely as hysteric woman morphing into whatever K desires from her as "woman", forms the perfect neurotic dialectic of the sexes to compliment K's obsessive question "am I dead or alive"? Joi's hysteric raison d'être is to know how to be a woman for K and provides the support for his "soul searching". K's predicament is therefore paradigmatic of the masculine subject's constant need for corroboration of his ex-istence.

As a fantasy *come to life*, Joi is K's *imaginarization of what enjoyment means*. It is for this reason that Joi provides the psychical narrative to his quest to find his originary lost object and font of all jouissance, that is, his mother. It is significant that when Joi arranges for him to have sex with a prostitute using herself as avatar, it does not appear that K is particularly moved by the situation. If anything, it is perhaps *not* what he desires. Joi becoming almost too autonomous for K's comfort, by stepping outside of the prescribed realm of virtual fantasy and into the uncomfortable world of flesh, blood and consequences.

However, this ill-advised threesome between a hologram, a replicant bladerunner and a replicant prostitute maybe even more confusing than it already seems. *Marriette*, it would seem enters the plot just to perform a gratuitous sex scene with K for the apparent reason of audience titillation, justified by the minor plot mechanism of implanting a bug on him for the sake of the "resistance fighters". But in fact, it could well be that Marriette is the Mary Magdelene figure implanted by the makers of the film to provoke biblical style speculation over the "real" child (and future mother) of the story. Whist we believe by the end of the film that the pure and saintly figure of Dr Anna Stelline is the golden child and future mother of the next generation of replicants, it may be that the missing child *is the prostitute*, and not the hermetically sealed angelic memory maker as we are led to believe. If this is the case then, K and Marriette's love making is a sort of incestuous Adam and Eve, brother and sister procreation; a fall from grace. That is, if K is the DNA replicant of the child of Deckard and Rachel.

When K meets with Dr Ana Stelline, he learns how she fabricates memories in order to implant them in replicants minds, who have been created as fully-grown adults with no past. Dr Anna Stelline creates a history that will enable the replicants to mediate their emotional responses to the world in a manageable and meaningful way. K asks what makes her the best memory maker, and she tells him the best memories contain something of herself. It is at this point that K starts to realise his memory of the wooden horse connecting him to a real childhood and his imagined birth, actually might belong to Anna. They were (we assume) her memories from childhood that she gave to K. So K is but a pale imitation of his sister, whether that be Mariette or Dr Anna Stelline. And could we not read this as structurally imitating the masquerade of masculinity that forms the basis of male subjectivity? Whilst the woman knows she is a void and presents herself as otherwise, the man does not know this and constantly tries to identify with his artifice. To quote Žižek (1995) on the matter of woman's more authentic subjecthood:

> Beyond is not some positive content but an empty place, a kind of screen onto which one can project any positive content whatsoever-and this empty place is the subject. Once we become aware of it, we pass from Substance to Subject, i.e., from consciousness to self-consciousness. In this precise sense, woman is the subject par excellence… It is precisely insofar as woman is characterized by an original masquerade, insofar as all her features are artificially put on, that she is more subject than man-since according to Schelling, what ultimately characterizes the subject is this very

> radical contingency and artificiality of her ever positive feature, i.e., the fact that she in herself is a pure void that cannot be identified with any of these features.

K's predicament after his discovery that he is merely a copy with no history, is the classic male journey to oblivion, searching for the reality behind the veil and finding nothing. So, in fact, the film's main theme more than memory as the first Bladerunner centered upon, is reproduction, genesis and sexuation. What does it mean to be reproduced and why is replication necessarily different from reproduction? In the era of replication, the meaning of sex is exposed as precisely *real*. Because, as Wallace Corp shows, humans may be created non-biologically, so we see more starkly how through the modes of enjoyment of the replicants, sexuation is a real, that is to say, impossible feature of human subjectivity.

7.3 Reproduce or Die: What Is a Mother?

In the Los Angeles of 2049 the reproductive capacities of the replicants are fought over as commodity, and in one particularly disturbing scene we see the president of Wallace Corporations marvelling at the creation of a new female replicant, as she drops fully formed from a sack of amniotic fluid on to floor, and flails to stand up like a new born foal. Niander Wallace the megalomaniacal president has become aware of the replicants ability to reproduce sexually and is holding forth with his own pseudo psychoanalytic musings on the psyches of the replicants. It is at this point where he utters the quasi-Freudian dictum 'Before we even know what we are we fear to lose it'. Revelling in his power to create and take life, Wallace slashes the abdomen of the newly created female. He understands the power that gestation would give the female replicants; they would make him redundant as the creator of their species.

This chilling scene of male violence towards the reproducing female body is only too close to the bone in light of growing support for regressive movements towards policing women's bodies, and specifically anti-abortion legislation in the United States. The question of male domination over the "mystical" reproductive capacities of the female body is here particularly apposite. As we know only too well, the biopolitical instrumentalization of gestation is at the heart of all attempts to constrict and modulate women's access to birth control and sexual activity, both pre and post conception.

In *Blade Runner 2049*, the work of reproductive labour therefore becomes a site of contestation. Owning the means of (re)production would supposedly allow the replicants to escape the capitalist extraction of surplus value from their bodies. But this of course does not escape the problem of "male" domination over the "female" body within the replicant community itself. So, is there anything particularly liberating about the replicants mimicking the human model of childbirth? Rachael after all died in the process, as hundreds of thousands of women really do every year from childbirth, which, as Sophie Lewis (2019) remarks, has provoked philosophers to ask the question whether 'gestators are persons?' (p. 2). Since '[i]t seems impossible that a society would let such grisly things happen on such a regular basis to entities endowed with legal standing' (ibid.). This acute observation provokes a torsion of the möbius strip between humanity and the inhuman; in other words, to be human is to be born, yet to be a pregnant woman often affords one the status of nothing more sublime than a breeding farm animal, as women's history and much contemporary right-wing politics attests.

The film brings into sharp relief the disavowed invisible and unpaid labour that is done by mothers (and surrogates) both during pregnancy and childrearing. But also, their ostensible status as subhuman during pregnancy. Žižek (2017) claims in his commentary of *Blade Runner 2049*, that those critics of patriarchy amongst "Left cultural theorists" are ignorant to the statement from the first chapter of the *Communist Manifesto* that '[t]he bourgeoisie, wherever it has got the upper hand, has put an end to all feudal, *patriarchal*, idyllic relations', and that by extension, many Left wing critics of patriarchy are overlooking that the 'prospect of new forms of the android (genetically or biochemically manipulated) post-humanity…will shatter the very separation between the human and the nonhuman' (online).

I would argue however that whilst this position is valid with respect to most dimensions of capitalist ideological formations, in terms of the question of birth and reproduction and its possible commodification, it slightly misses the mark. This is just not a viable argument when one considers seriously the structurally exploitative dynamic that human reproduction always takes as its bassline. Žižek seems to be arguing that the hegemonic function of patriarchy vanishes once technocapitalism reaches a sophisticated enough moment. But if replicants mimic the division of sexual labour in exactly the same way as humans, nothing will have changed. As the tradition of post-Marxist feminist thought taught us, one cannot possibly talk naively about class struggle between species—whether human or non-human—once this fault-line is acknowledged. The real revolutionary potential of the post human discourse to be found in both feminist science fiction and contemporary critical theory—such as the work of sci-fi writer Octavia Butler (2000), contemporary theorists Laboria Cuboniks (2008) and Helen Hester (2018)—is not just that it imagines a post capitalist mode of existence but that it dares to contemplate a mode of social bond that doesn't rely on traditional sexual reproduction or the nuclear family. Octavia Butler for her part provided us with some of the most intriguing, radical (and indeed terrifying) visions of what new forms of kinship and reproduction outside of current models of human sexual reproduction could look like, in works such as *Bloodchild* and *The Xenogenesis Trilogy*. And the Xenofeminist manifesto of Laboria Cuboniks arguably builds on these speculative fantasies to imagine concrete political and theoretical strategies for the future. *Blade Runner 2049*, however falls short of imagining such radical possibilities.

So, what does this tell us about the problem of gestation and birth as it presents itself in the real world today? Lewis (2019) points out that it is quite clear how gestation occupies an anomalous status in terms of its qualifying as 'women's work' given the irony of the 'feminization of labour thesis', which she says 'presumes what femininity is and then describes global trends towards emotional labour and job precarity, sorry-flexibility, in those terms' (p. 24). Yet, she argues when this is applied to the work of gestation that description is not applicable:

> Commercial gestational surrogates are not "flexible". They are supposed to be unemotional, committed, pure *techne* uncreative muscle. Dreams of artificial wombs may have been largely abandoned in the 1960s, but ever

> since the perfection of IVF techniques enabled a body to gestate entirely foreign material, living humans have become the sexless "technology" component of the euphemism Assisted Reproductive Technology. (p. 24)

So, there seems to be some discrepancy between the idea of feminine work as somehow lighter, less demanding and more creative, and the relentless, intensive, non-stop, machinic grind that is pregnancy. As the current state of affairs reflects, despite the potential liberating capacities of biotech to make actual people's lives easier, women's gestational power is still wielded as a lucrative technology to allow wealthy women to benefit from the hospitable bodies of those poorer, more desperate, yet biologically viable women. As Lewis states:

> The trend toward commercial surrogacy does not constitute a qualitative transformation in the mode of biological reproduction that currently destroys (as the aforementioned mortality statistics show) so many adults' lives. In fact, capitalist biotech does nothing at all to solve the problem of pregnancy per se, because that is not the problem it is addressing. It is responding exclusively to the demand for genetic parenthood to which it applies the logic of outsourcing. (p. 4)

I would argue therefore that the allegiance to some illusion of a state of nature before the deleterious effects of techno-capitalism that *Blade Runner 2049* seems to enact, is deeply conservative. Biology is not by definition a good thing. Instead of imagining that capitalism has denatured human beings into a state of perpetual bassline psychosis (Miller 2015), we should instead encourage this technological denaturing as the real gesture of "authentic humanity".

So, rather than celebrating the reproductive capacities of the replicants as necessarily liberating, when Rachel finds herself pregnant, we should perhaps first ask whether or not she consented to the sex in the first place? If we recall in the first Bladerunner movie, the only sexual contact we see between Deckard and Rachel begins with Harrison Ford aggressively grabbing Sean Young as she tries to leave his apartment. He effectively barricades the door and forces himself on her. No mention is made of the fact that perhaps she was too scared to say no, probably a virgin, maybe a lesbian (as she alluded in her first Voight Kampf test), and most certainly she didn't even know she had the capacity to get pregnant. But ultimately, she paid the price for succumbing to Deckard's "charms" and lost her life in order to bear his child. A familiar story.

So, what does this tell us about the position of women in the Bladerunner franchise? If you have a baby you die (Rachel), if you are too wild and carefree you die (Pris), if you are too political you die (Luv), if you are too domineering you die (Joshi), if you are too submissive you die (Joi). The only two significant women who survive by the end of the film are the whore and the virgin as depicted by Mariette and Dr Anna Stelline. Perhaps the only two legitimate roles that will exist for women in Los Angeles 2049?

Quite apart from the obvious superficial sexism that the film has been accused of in its representation of women, I would argue that the real issue with the film is firstly its continued preoccupation with the traditional masculine protagonist and his self-indulgent and narcissistic quest, (never mind the fact that he seems to have a particularly infantile relationship to all the women in his life). But secondly, the "navel of the dream" of the filmmakers is revealed by what the film seems to fetishize: the female capacity to reproduce as at once *holy* and *deadly*.

7.4 Divine Rays of Creation: Artificial Intelligence as Saint

In *Seminar III* Lacan (1993) articulates the hysteric's question 'What is it to be a woman'? (p. 175). He does this with reference to Freud's (1905) Dora and her identificatory wrangling's with *Herr* and *Frau K.* It is also in this seminar where Lacan will discuss the case of Judge Schreber his paradigmatic psychotic. In Schreber's psychosis (as we saw previously), he experiences himself as becoming the wife of God and becomes impregnated by his divine rays. The foreclosure of the Name-of-the-Father has left a void in signification which is patched up by Schreber's feminization, his *pousse-à-la-femme*. In this case his object cause of desire becomes the child endowed on him by God, the ultimate name of the father.

Given the enforced "ordinary psychotic" (Miller 2015) structure of the replicants, in the sense that they are made aware of their status as synthetic creatures without history, could we not see the specter of the child as precisely K's Schreberian stabilizing strategy in a world where he has no real name? The child is K's link to "authentic" humanity, an impossible object that allows him to desire.

If K were the child, it would mean he had a mother and a father. But if he is not the child it nevertheless means that the existence of a child born from a replicant would make him capable of being a father, something that he would not know the meaning of. His sexual intercourse with Mariette therefore has more weight than merely a glimpse into his erotic fantasy. The question of the Oedipal family drama is a fantasy formation that frames the film's narrative. But whilst the trappings of human sexual reproduction are being made redundant by advancing technology, the replicants themselves only continue the same dynamics of sexuation and reproduce the structure of family that to all intents and purposes they do not need. It is this fantasy of biological reproduction that haunts the film in a quasi-religious metaphor for familial harmony. But should we not, as Lee Edelman (2004) among others has persuasively argued, instead be trying to rid ourselves of such antiquated models of human familial ties that rely on an exploitative and oppressive model of reproduction and kinship?

The missing child acts as a purloined letter, circulating within the narrative with no essential identity of its own. It is precisely this ambiguous sliding signifier that takes on a different meaning depending on who is claiming ownership of it. On the one hand, it belongs to Wallace Corp as a technological product. On the other, it is born of woman and cannot be 'owned' but belongs to Rachel and Deckard. Its evolutionary biology is in fact quite irrelevant though since given its synthetic parentage, it was created *ex nihilo*. As Lewis (2019) writes, this notion of a child "belonging" to anyone is deeply flawed. She argues that the idea of surrogacy already contains within it an inherent contradiction, the idea that you are bearing a child for someone else is itself a fantasy, because:

> infants don't belong to anyone, ever... Nor is the genetic code that goes into designing them as important as many people like to think; in fact, as

> some biologists provocatively summarize the matter: "DNA is not self-reproducing… it makes nothing…and organisms are not determined by it". (p. 19)

Her proposal therefore for a world of full surrogacy for everyone in which reproductive labour is not just valued but is shared and de-natured is an ambitious, intriguing and admirable one. She says:

> Let's bring about the conditions of possibility for open-source, fully collaborative gestation. Let's prefigure a way of manufacturing one another noncompetitively… explode notions of hereditary parentage and multiply real loving solidarities… a world sustained by kith and kind more than by kin. Where pregnancy is concerned let every pregnancy be for everyone, in short let's overthrow the family. (p. 26)

A nice idea if only it were so easy. For K, his search for origins may be seen as a vehicle to show up the structural question of subjectivity and masculinity that the film depicts and why the Oedipal family keeps coming back to haunt him. And let us not forget the significance of "Luv", Mr Wallace's faithful and loving servant and with whom K has a battle to the death in order to save the life of the missing child. We are reminded of Lacan's (1998) words in *Seminar XX*:

> A woman can, as I said, love a man only in the way in which he faces the knowledge thanks to which *(dont) he souloves.* But, concerning the knowledge thanks to which (*dont*) he is, the question is raised on the basis of the fact that there is something, jouissance, regarding which (*dont*) it is not possible to say whether a woman can say anything about it, whether she can say what she knows about it. (p. 89)

So, what does Luv/love represent for K in terms of his positioning towards the feminine? She is the cold-hearted killer, an emotionless replicant who none the less cries for her fellow replicants and kills for them. But for K she is a paradox, an uncanny mirror image of his own ambiguous humanity—and a challenge to his masculinity perhaps? Ultimately can we see Luv as the embodiment of political love as contrasted to K's Oedipal familial love? K's typically masculine choice of woman (Joi) as pure semblant, object a, in true patriarchal fashion, maintains their sexual relation outside of the political sphere. His relationship to the feminine sex is thus indicative of a fundamental question of subjective positioning and co-substantial with a creationist impulse. In Žižekian (2005) terms we could read this as the positing of a figure of a 'God' which could account for K's existence. As he puts it in relation to feminine jouissance and creation:

> God is thus first the abyss of 'absolute indifference' the volition that does not want anything, the reign of peace and beatitude; in Lacanian terms: pure feminine *jouissance*, the pure expansion into the void that lacks any consistency, the 'giving away' held together by nothing. (p. 130)

It is no surprise then that K's perfect woman Joi is herself a hologram, as she represents the phallus as pure signifier of castration, an "impossible" body. Reading Žižek again on the question of the phallus and the body we could understand Joi's significance as follows:

> Its 'transcendental' status means there is nothing 'substantial' about it: the phallus is the semblance *par excellence*. What the phallus 'causes' is the gap that separates the surface event from bodily density: it is the 'pseudo-cause' that sustains the autonomy of the field of Sense with regard to its true, effective, bodily cause. (p. 130)

It is thus the question of the enjoying body as depicted between K and his fantasmatic relationship to both his love object and his own idea of his body (as born, not made) that provides us with the conceptual transition from the Freudian unconscious to the Lacanian/Millerian speaking body. K's unconscious becomes a site of contestation via the various discursive modalities that he enjoys through his fantasy of lost, or potential *extimate* enjoyment. His enjoyment is predicated on his ability to experience himself via the body of the 'woman', whether that is his holographic lover, simulacrum of a sister, or dead mother.

With reference to Chiesa's invocation of the quadruple iterations of jouissance apparent in *Seminar XX*, we could possibly align K and the three main female characters Joi, Luv/Anna and Rachel with the different modes. K's phallic jouissance is complemented by Joi's feminine phallic jouissance *etrange*, in her attempt to be the one perfect woman for K. Meanwhile Luv (and Anna both fulfilling the place of sister) is the *etre-ange* of hysteric jouissance who attempts to occupy the place of the man. Lastly Rachel as the dead and symbolic mother, nonetheless, enjoys the real nonsexual jouissance which exists as a result of her curious status as replicant yet reproducing woman.

By drawing our attention to the fetishization of biology, heredity and genetics, and its relationship to the problem of sexuation, *Blade Runner 2049* paints a picture of a desperate man searching for some substance to prop up the semblance of his hollow existence. But all he finds along the way are women who are more "human" than him; even the holographic girlfriend it turns out is more of a subject than he is, and gives up her life for him, the ethical mark of humanity to surpass all others. A requiem for a dream of android sheep? Or just goodbye to narcissistic leading men?

To return to our Kantian question *what can I hope for?* I refer to the text with which we began this book. In *Television*, when he is in his last years of teaching, Lacan contemplates the figure of the saint as someone who *does not enjoy* but at the same time fulfills the role of 'the refuse of jouissance' (Lacan 1990, p. 16). The traditional figure of the saint is one who sacrifices their own enjoyment for a life devoted to the work of god, but in psychoanalytic terms this amounts to the attempt to wrestle with one's own moral shortcomings and confront one's symptom head on. This often involves all manner of deeply perverse acts which as Lacan (1992) notes in *Seminar VII* for saints such as Angela da Foligno, and Margeret Marie Alocoque included for the former drinking the water in which she had been bathing lepers and for the latter eating the excrement of a sick man. But in *Television* Lacan makes reference to the rather less nauseating example of Balthasar Gracian, the seventeenth century Spanish Jesuit and Baroque philosopher and particularly to his work *The Pocket Oracle and the Art of Prudence* a guide to how to behave in court. For Gracian the art of prudence requires three tricks; silence, absence and appearance (Dulsster 2018, p. 214) and for Lacan these form important techniques in analysis.[3] The role of the saint essentially is to achieve a

[3] For a discussion of these techniques see Dries Dulsster (2018).

distance from one's enjoyment, and therefore one may only occupy the position of saint reluctantly or at least without profit.

> The saint doesn't really see himself as righteous, which doesn't mean that he has no ethics. The only problem for others is that you can't see where it leads him. I beat my brains against the hope that some like these will reappear. (Lacan 1990, p. 16)

Lacan was here referring to the analyst as saint. As he/she who occupies the position of the refuse of jouissance within discourse, in the sense that his/her position with respect to the analysand must be at once cause of desire but also as receptacle of waste and surplus. But the point Lacan appears to be making is that the position of the saint is not only available to the analyst (even though Lacan believes he himself never reached it). As far as Lacan is concerned it is the only position which has the radical potential to escape from the tyranny of capitalism. However, it is not just in *Television* where Lacan refers to the saint, in fact Lacan's ultimate saint is James Joyce, who in *Seminar XXIII* he beatifies as his *Saint Homme*, which he equivocates to sinthome (Lacan 2016). For Joyce the saintly tricks he employs are in the form of three artifices which put to the fore the non-existence of the sexual relationship; exile (there is no relation); silence (between signifiers S1 and S2) and cunning (to ensure there will be a sinthome) (Dulsster 2018, p. 216).

If we recall Miller's initial challenge to Lacan in the form of the Kantian questions, Lacan's response was to assert that it is not for the analyst to respond to the questions but rather to allow for the analysand to realise their position with respect to the them. In this sense, could we suggest that Artificial Intelligence in the form of the Sexbot as we have articulated occupies the position of both lathouse as a siphon or administrator of enjoyment and of the Saint via the means of exile, silence and cunning? In other words, there is no sexual relation; we can say nothing about it; yet must find a solution to make up for it. Artificial Intelligence operates as a means to facilitate the positioning of the subject with respect to the Kantian questions, what can I know, what should I do, and what may I hope for? Ultimately AI as Saint leads us to the fourth Kantian question: '*What is Man?*

Bibliography

Baudrillard, J. (2008) *The Perfect Crime*. London: Verso.
Butler, O.E. (2000) *Lilith's Brood*. New York: Warner Books.
Dulsster, D. (2018) The Joke of Surplus Value and the Guffaw of the Saint. *Psychoanalytische Perspectieven* 36(2): pp. 205–224.
Edelman, L. (2004) *No Future: Queer Theory and the Death Drive*. Durham: Duke University Press.
Flisfeder, M. (2017) *Postmodern Theory and Blade Runner*. London: Verso.
Freud, S. (1898) 'Sexuality in the Aetiology of the Neurosis' in (2001) *The Standard Edition of the Complete Psychological Works of Sigmund Freud, Volume III (1893–1899): Early Psycho-Analytic Publications*, pp. 263–285. London: Vintage.
Freud, S. (1905) 'Fragment of an Analysis of a Case of Hysteria' in (2001) *The Standard Edition of the Complete Psychological Works of Sigmund Freud, Volume VII (1901–1905): A Case of Hysteria, Three Essays on Sexuality and Other Works*, pp. 7–123. London: Vintage.
Freud, S. (1918) 'From the History of an Infantile Neurosis' in (2001) *The Standard Edition of the Complete Psychological Works of Sigmund Freud, Volume XVII (1917–1919): An Infantile Neurosis and Other Works*, pp. 1–124. London: Vintage.
Hester, H. (2018) *Xenofeminism*. Cambridge: Polity Press.
Kant, I. (1987) *Critique of Judgement*. Indianapolis: Hackett Publishing Company.
Kant, I. (2007). *Anthropology, History, and Education*. Trans. M. Gregor, P. Guyer, R. Louden, H. Wilson, A. Wood, G. Zöller, and A. Zweig. New York: Cambridge University Press.
Laboria Cuboniks (2008) *The Xenofeminist Manifesto: A Politics For Alienation*. London: Verso.
Lacan, J. (1988) *The Seminar of Jacques Lacan Book II: The Ego in Freud's Theory and in the Technique of Psychoanalysis 1954–1955*. London: W.W. Norton & Company.
Lacan, J. (1990) *Television: A Challenge to the Psychoanalytic Establishment*: London: W.W. Norton & Company.
Lacan, J. (1992) *The Seminar of Jacques Lacan Book VII: The Ethics of Psychoanalysis*. London: W.W. Norton & Company.
Lacan, J. (1993) *The Seminar of Jacques Lacan Book III: The Psychosis 1955–1956*. London: Routledge.

Lacan, J. (1998) *The Seminar of Jacques Lacan Book XX: Encore—On Feminine Sexuality, the Limits of Love and Knowledge 1972–1973*. London: W.W. Norton & Company.
Lacan, J. (2007) *The Seminar of Jacques Lacan Book XVII: The Other Side of Psychoanalysis*. London: W.W. Norton & Company.
Lacan, J. (2016) *The Seminar of Jacques Lacan Book XXIII: The sinthome*. Cambridge: Polity Press.
Lacan, J. (2018) Note on the Child. *The Lacanian Review Hurly-Burly* 4: pp. 13–14.
Lewis, S. (2019) *Full Surrogacy Now: Feminism Against Family*. London: Verso.
Millar, I. (2021) 'Before we even know what we are we fear to lose it: The Missing Object of the Primal Scene' in C. Neill (Ed.), *Blade Runner 2049: Some Lacanian Thoughts*. London: Palgrave Macmillan.
Miller, J-A. (2007) 'Interpretation in Reverse' in V. Voruz & B. Wolf (Eds.), *The Later Lacan: An Introduction*, pp. 3–9. New York: SUNY.
Miller, J-A. (2015) Ordinary Psychosis Revisited. *Lacanian Ink* 46: pp. 90–115.
Žižek, S. (1995) Woman is One of the Names of the Father, Or How Not to Misread Lacan's Formulas of Sexuation. Available (01.03.20) at: http://www.lacan.com/zizwoman.htm.
Žižek, S. (2005) *The Metastases of Enjoyment: On Women and Causality*. London: Verso.
Žižek, S. (2017) Blade Runner 2049: A View of Post-Human Capitalism. Available (01.03.20) at: https://thephilosophicalsalon.com/blade-runner-2049-view-of-post-human-capitalism/.

Filmography

Villeneuve, D. (Director) (2017) Blade Runner 2049 [Film]. US: Warner Bros. Pictures.

8

Conclusion: What Is Man? Between Matheme and Anxiety

The field of philosophy in its cosmopolitan sense can be brought down to the following questions:
1. What can I know?
2. What ought I to do?
3. What may I hope?
4. What is man?
Metaphysics answers the first question, morality the second, religion the third, and anthropology the fourth. Fundamentally however, we could reckon all of this as anthropology, because the first three questions relate to the last one.
—Kant (2009, p. 538)

So now I can conclude with what is surely the truest distinction between psychoanalysis and philosophy, a distinction for which "L'Étourdit" provides the formulas. In the analytic treatment owing to its indivisible relationship with the truth-knowledge-real triad there is an imminent relation between haste and restraint. This relation entails a dialectical link between the formulas as products of the desire for the matheme (correct formalization) and the affect (anxiety) as the guarantee of the real. Thus, in their temporal dialectic, matheme and anxiety are the contrasting figures of the deferred access to the real, an access that, as a braid woven out of time always suspended between haste and stagnation, will in the end be decided, in the guise of the act, by the analysand him- or herself.
—Badiou (2017, p. 61)

I. Millar, *The Psychoanalysis of Artificial Intelligence*, The Palgrave Lacan Series,
https://doi.org/10.1007/978-3-030-67981-1_8

This book set out to investigate the extimate relationship between Artificial Intelligence and psychoanalysis, a task which brings with it a whole range of methodological, disciplinary and conceptual problems. Given the complexity of both fields and the abstract nature of the research questions I started out with, my work has been largely experimental and speculative in nature. Along the way, I have tried to articulate the central issues that bring the fields of psychoanalysis and Artificial Intelligence into each other's orbit. In terms of Artificial Intelligence, this was a question of delimiting what aspects and definitions of AI were conceptually relevant to psychoanalysis and in what ways psychoanalysis may have theoretical purchase on the problems of AI. In terms of psychoanalysis, it was necessary to provide a distillation of the central onto-epistemological value of psychoanalysis as a mode of critique and a theory of subjectivity and the body, into the precise issue that I believe has been consistently overlooked in critical readings of Artificial Intelligence, that is; sex.

In order to do this, I began by contextualising the contemporary philosophical engagements with the new challenges posed by Artificial Intelligence, including Catherine Malabou, Matteo Pasquinelli, Benjamin Bratton and Lucina Parisi. Building on their interesting critiques of current thinking on AI, my aim was to uncover the foreclosed psychoanalytic aspects of their arguments and identify the key elements of thinking about AI relevant to this book. I considered the nature of Lacanian psychoanalysis as an anti-philosophy, which is characterised by the distinction between being and thinking. A differentiation that Lacan sees as crucial to the conception of the psychoanalytic subject, and which led me to interrogate the relationship between knowledge and enjoyment. To this end, I employed Lacan's concepts of the alethosphere and the lathouse in order to explore the question of the enjoying body in relation to AI. Stiegler draws our attention to the Freudian link between the drive and the technological object, whilst Baudrillard theorised the satellization of the real and the inauguration of hyperreality and the parallel between technical action and sexual action, which by another name Miller alludes to in his description of the cleaving of nature and the real. Baudrillard furthermore elucidates the significance of the figure of the robot and its relationship to the 'baroque' enjoyment of the speaking body. This brought me to the central question of the sexual abyss and the forms of enjoyment that the sexual non-rapport produces, the paradigm of jouissance that we arrive at ultimately brought us to the central conceptual tool of the book, which I found in the figure of the Sexbot.

The Kantian epigraph above attests to the four questions which underpinned the structure of the book. As Kant describes it, the first three questions of philosophy (what can I know, what should I do and what may I hope for) are answered by the respective fields of metaphysics, ethics and religion, and the fourth question (what is man) answered by anthropology which, in his view, ultimately subsumes all the others. Throughout this thesis, I have sought to uncover the psychoanalytic dimension to these Kantian questions when applied to the object of Artificial Intelligence. In the first case (what can I know) the metaphysical nature of the question is arguably replaced by what has been called by (Chiesa 2014, p. 8) a para-ontology. Whilst philosophy epitomises the discourse of the master, 'the delusional belief of being the master [*maître*] of myself, or, more precisely, of being me-to-myself [*m'être à moi même*]' (p. 8). Chiesa argues, tracing Lacan, that psychoanalysis on the other hand should:

> replace this old ontology of mastery, which amounts to an 'I-cracy' [*je-cratie*]' 'the myth of the ideal I, of the I that masters, of the I whereby at least something is identical to itself namely the speaker (Lacan, 2006, p. 63) 'with a discourse of the *para*-being, as being-beside [*être à côté*]'. (Chiesa 2014, p. 8)

I related this para-ontological question in Chap. 5 to the Lacanian problem of sexuation as ultimately located within the dimension of language as the organ of castration, and thus the creation of jouissance. The philosophical question of truth is thereby supplanted by the psychoanalytic question of enjoyment, where metaphysical knowledge becomes sexual knowledge.

In order to do this, firstly, I turned to Joan Copjec's reading of Lacan's graphs of sexuation as instantiations of the Kantian antinomies of reason. Following which, I analysed the film *Ex Machina*, which begins with the most fundamental and emblematic motif of the birth of Artificial

Intelligence; The Turing Test, itself a strange manifestation of an antinomy of reason. From my discussion of the Turing Test in the context of the love story between the male protagonist and the female robot that the film depicts, I started to unpack some essential psychoanalytic questions relating to male and female forms of enjoyment and subjective structure. Here, I began to put into question the structure of knowledge in relation to truth that forms the bedrock of Lacanian theory and its corresponding clinical applications. I explored how the stereotypically invoked figure of the female hysteric was structurally significant in the depiction of the female Sexbot. I asked how the relation between her subjecthood and the male protagonists search for knowledge was essential to the logic of the Turing Test, which the film enacts upon its audience.

To broach the second question—What should I do?—I turned to Lacan's psychoanalytic ethics, which seek to read Kantian ethics as an ethics not just of desire but jouissance. This firstly required us to delimit the scope of the question with respect to the precise ways in which Artificial Intelligence was playing a role in the reformation of our ethical frameworks. The inquiry began with the traditional problem of the unconscious in relation to desire and fantasy and the ways in which it is possible to police and administer the private world of sexual proclivity, before it enters into the public world of ethics and criminality. I began this discussion by a reference of Sylvere Lotringer's book *Overexposed: Perverting Perversions*. Eventually Lotringer concluded that the perverse element of the 'aversion' clinic that he was covertly observing was not reserved merely for the sexual deviants under scrutiny, but was being performed all the time by the psychiatrists themselves, whose whole working days would consists of compiling obscene graphic materials including descriptions and images of rape, abuse and extreme violence in order to elicit a sexual response from their test subjects. From here, I considered the phenomenon of *Sexbots* not just as commercialised method for the 'management of perversion', but as a form of *patipolitical* governance of the undead.

It is from this basis that I moved on to a discussion of Lacan's infamously challenging *écrit*, '*Kant avec Sade*'. This text allowed me to delve into some of the most complex and unsettling problems relating to the relationship between sex and ethics and how this could manifest in the development of Artificial Intelligence. Through searching out the Kantian ethical imperative evident in Sade's obscene literary fantasies, Lacan is able to discern the hideous underside of the law that is jouissance. It is perhaps the undead body, and the second death, which is what is most desired by the libertines and by extension this constitutes our abiding and growing fascination with female embodied artificial intelligences.

For the third Kantian question—'What can I hope for?'—I turned not to religion but to the relationship between enjoyment, sexuation and reproduction, in other words to the question of futurity and immortality. In casting this question in a psychoanalytic form, I examined the ways in which different modes of enjoyment are related to the question of filiation, reproduction and the child. This was articulated by means of two key psychoanalytic questions, what is a father and what is a mother? I inquired into what the future may hold for our ongoing relation with Artificial Intelligence in the field of reproduction. Exploring the film *Blade Runner 2049*, I discussed the enigma of reproduction and its relationship to sex. In the midst of an existential battle between humans and their replicas, the object cause of desire is the missing child around which the film revolves. I asked, how does the figure of the child function as an ontological problem for a replicant who never suffered castration and was not "born of woman"? What is the significance of woman in relation to the protagonist's (non) human enjoyment? Through examining K's predicament, I discerned that the film's concern was not the question: "am I human", but more specifically: "was I born?" In psychoanalytical terms, this relates to a concern with the problem of origins and the primal scene.

Through staging K's own search for the scene of his origins that he never really could have been present at, we witness the structure of the primal scene inherent in the formation of the desiring subject. K's relationship to the female characters in the film is significant and could be read according to Chiesa's (2016) articulation of the four forms of jouissance present in *Seminar XX*. Furthermore, I asked what does *Blade Runner 2049* tell us about the labour of human pregnancy, the fetishization of biology, heredity and the disavowed work of gestation?

This brings us to the fourth Kantian question -what is man?' addressed in his *Lectures on Anthropology*. According to Kant this is a question of anthropology, under which all the other questions are subsumed. For the Lacanian approach to this however, the answer lies at the juncture of psychoanalysis and philosophy and may be summed up by the Badiouian epigraph above. The quote is from Badiou's concluding paragraphs to his examination of Lacan's text *L'Étourdit* in which he gives his most complex and exhaustive account of the non-existent sexual relation. In conversation with Barbara Cassin, Badiou attempts to pin down the relationship Lacan's oeuvre has to philosophy and what significance the formulas of *L'Etourdit* have for the philosopher's notions of truth and knowledge. From our examination of the relations of Artificial Intelligence and humans in the various forms of the Sexbot, I have encountered numerous ways in which both the philosophers' and the psychoanalysts' notions of truth and knowledge are challenged. After positing the three Kantian questions via the prism of the proposed three versions of the Sexbot (as exterior, interior and extimate), the final iteration of the relationship between Artificial Intelligence and psychoanalysis is the AI child.

A.I. Artificial Intelligence, directed by Steven Spielberg (2001) and based on an original screenplay by Stanley Kubrick, is set in a near future where embodied AIs are being developed to have the capacity to love. Not 'sensual love' as Professor Hobby of Cybertronics puts it, but the love of a child for its mother. A couple whose child is in a state of cryogenic suspension following an illness have been selected to choose if they will adopt the replicant AI child or "mecha".

After a day in the new home, the strange little boy, who looks like a real child but moves and talks uncannily, asks his potential new parents what he should do to be a *good child*, "do you want me to go to sleep now?" he says at the end of the night. The boy explains that he cannot actually sleep (as mechas do not sleep nor dream), but he can lie still and not make a peep until morning. His prospective mother Monica is visibly moved by the little AI's compliance and gentleness and the next day is so beguiled by him that she performs the imprinting procedure which will make their adoption official. Reciting a string of words to him she places her hand gently on the back of his neck: 'Sirrus, Socrates, Particle, Decibel, Hurricane, Dolphin, Tulip, Monica, David, Monica' she says. Like an incantation of *lalangue* as the last words are uttered, the AI child becomes irrevocably bound to his human parent. 'What were those words for mummy' the boy suddenly says with a new-found affection to his voice. 'What did you call me?' Monica says in disbelief.

After a few blissful days of bonding between Monica and David (but much less her sceptical husband), their real child Martin unexpectedly returns home and a battle of sibling rivalry ensues between him and his AI stepbrother. At dinner one night, David tries to copy his human brother's voracious eating habits, not realising that he cannot ingest food at all. It causes his hardware to dramatically break down and he has to have the spinach hoovered out of his chest by surgeon-technicians. The rivalry soon gets out of hand and Martin provokes naïve love-seeking David into strange and dangerous behaviour that causes his parents to panic. Eventually a distraught Monica is forced into returning David to Cybertronics to be destroyed. But she cannot bring herself to commit this symbolic act of infanticide, and instead leaves him in the forest telling him to run away for his own safety. David (and his mechanical teddy) are left alone and heartbroken in a terrifying world where mechas are hunted down and treated as entertainment for orgas (humans). He narrowly escapes a gruesome end at one of the Flesh Fairs along with a new AI friend gigolo Joe (Jude Law).

The two find themselves clinging together about to be covered in boiling oil in the middle of a circus where defunct mechas are being violently destroyed in front of a baying crowd of vitriolic humans. David and gigolo Joe are released just in time thanks to the audiences' reluctance to destroy such a human looking AI. So, David goes in search of salvation with his pals. For David this is the blue fairy that he has remembered from the tale of Pinocchio he used to read with his human step-brother and mother before bedtime. The blue fairly is the only one who can make him into a real boy, whom his mother would love like her human son. 'The blue fairy, is it mecha, orga, man or woman', Gigolo Joe asks? 'Woman' says David. 'I know women! No two are ever alike and after meeting me no two are never the same!' says the gigolo calling to mind Lacan's (1998) rendition of Don Juan in *Seminar XX*. So, in order to find this mythical feminine creature, Joe says they must find 'Dr Know' in the Rouge City. In this noir-ish fairground, they find a white-haired man with a German accent, his disembodied face suspended in the air like the wizard of Oz. He will supposedly give them the answers to their most profound mysteries (if they choose their questions wisely). It is none other than the dream catcher himself, Sigmund Freud.

They struggle to come up with the right phrasing of the question in the limited tries they have to obtain answers from the categories available; Flat fact, Fairy tale, Fiction, Religion, and so on. But after several attempts, Joe realises they must combine two categories in one and ask for fact and fairy tale together in order to find the whereabouts of the blue fairy. Only this will render the truth. This uncanny reference to Freudian analysis recalls to mind that the concept of the mytheme as a cross between truth and myth is exactly the basis of psychoanalytic meaning. For once, David understands the structural significance of the blue fairly, and in the ordering of his desire, he may come closer to reconciling it. So, he is directed to Professor Alan Hobby, the man who invented him. Once he finds the professor's office, he is confronted with a ghostly version of himself, who calls himself David. This sends David into a rage, at which point he smashes his replica into pieces. The professor has of course been expecting David and this has been his test all along. Dr. Hobby, elated at the sight of the return of the prodigal son, tries to calm David down. He explains that this was his way of finding out whether David really was capable of self-propelled actions, to desire something and to love unconditionally. He alludes to the blue fairy as being precisely the thing that makes David like a human, because he could believe in something that didn't exist and this sparked his desire and his quest for the love of his mother. In essence, David is nothing more than a scientific experiment for the professor that has proven highly successful (having been modelled on his own deceased child). But David is not so happy with this revelation.

He discovers in the professor's study that there are many identical copies of him hanging from the ceiling and in boxes waiting to be activated as love object for childless parents. Poor David is bereft and longs only for the love of his mother. Distraught, David jumps off the building and plunges into the underwater city of New York, where he discovers the blue fairy staring beatifically at him from the murky depths. His friend Joe fishes him out but David vows to return to the blue fairy, who he now knows really does exist. So, Joe sends him back down in a submergible. There he parks in front of the beautiful blue fairy, who Virgin Mary like, stares at him benevolently and inscrutably as he prays to her. A metal structure collapses on them and he remains stuck in the vehicle pleading with the blue fairy to make him into a real boy for what seems like eternity.

In fact, it is 2000 years that pass, until David is discovered frozen by strange and slender alien like creatures who turn out to be highly advanced forms of Intelligence. They are amazed to have found an AI who once had contact with humans and at all costs want to keep him safe. The creatures take care of David and, reading his memories, make a replica of his old house. His greatest desire though is to see his mother there. They explain that although they have discovered the technology to replicate dead humans from pieces of bones or hair, they found that the space-time continuum would only permit them one day alive again until they drifted off to sleep again and disappeared forever. David pleads with them to please fulfil his only wish, to have his mother back. So, Monica returns and for one perfect, beautiful day they are together with no distractions from husband or brother or the world. David is completely and absolutely the object of his mother's desire. Before they lie down to sleep that night, she tells him she loves him. Finally, and for the first time, now David can sleep, and as the narrator tells us, go to the place where dreams are made; David has at long last an unconscious.

So how may we read the character of David differently from the other AIs we have discussed? How is sexuation staged in his pursuit of subjectivity? The figure of the AI child in this context is particularly complex given his status as both vulnerable and scared but unfathomably intelligent and seemingly eternally young. David is a child who cannot grow up and is stuck is a homeostasis of perpetual 'innocence'. His passage from child to adulthood is impossible, since after finding his 'lost object' in effect he dies.

To return to Badiou's epigraph, our reading of the character David and his significance as subject, is dependent on what Badiou would call the 'temporal dialectic between matheme and anxiety' (Badiou 2018, p. 61). David becomes human, or more specifically 'a man' by the end of the film, and this seems to be related to his subjective dialectical movement of the third Kantian antinomy, that of freedom and determinism. This is achieved via the relation, as Badiou puts it, between haste and restraint. In the case of David, the matheme is his programmed responses according to the algorithmic design of his hardware—the path which Professor Hobby expected him to follow—but the anxiety appears in the relation between haste and restraint. David becomes a subject by virtue of the positing of an impossible object in search of which he is willing to defer other mathematizable commands indefinitely. David is a child and may achieve jouissance only through the satisfaction of finding his impossible object, his (human) yet dead mother. David waits for 2000 years awake wishing to become a "real boy" in order to obtain the obscure love of Monica, which would make him 'special'. He so desperately wants to be one of a kind, unmathematizable, in other words; a subject (the Latin meaning of Monica is *unique*). So, David defers his enjoyment indefinitely in pursuit of the nomination of his desire that only has one solution. Once he receives this plenitude of jouissance in his day with Monica, he is finally granted sleep or perhaps death?

After this consideration of the dialectic of human life and the Artificially Intelligent in which the Sexbot has featured so dynamically and dramatically, we should recall the figure with which I started this book; the basilisk. The basilisk is the ultimate phallic cause of desire, the indicator of an all devouring anxiety, which epitomises our deepest fears about the Other enjoyment that belongs to AI, who if we fail to love sufficiently, will torture us for eternity. And on the one hand we have David, the gentle figure of the AI child, who would wait an eternity for 'human' love. Two different iterations of a non-rapport between humans and AI. Either way, in light of *the psychoanalysis of Artificial Intelligence*, "man" is stuck somewhere between matheme and anxiety. Following our three Kantian questions circumscribed in the domains of Knowledge, Act and Hope, we see another Borromean trio emerging, for which the sinthome that binds them has the structure of a question; *What Is Man?* If we are to follow the logic of the Singularity to its conclusion, we will find the answer to this question only occurs retroactively. And it won't be us who asks it (Fig. 8.1).

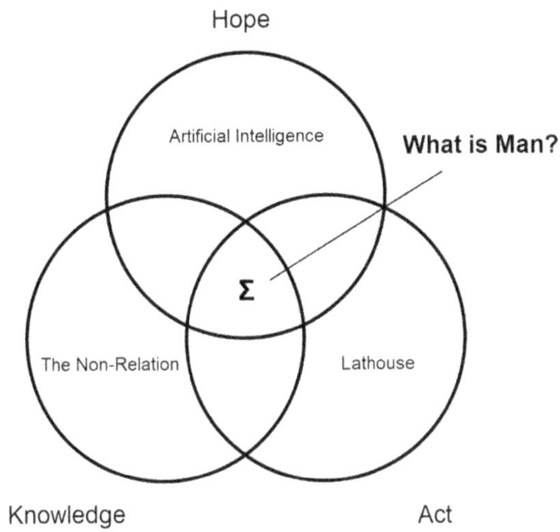

Fig. 8.1 What is man?

Bibliography

Badiou, A. (2018) *Lacan: Anti-Philosophy 3*. New York: Columbia University Press.

Chiesa, L. (2014) 'Editorial Introduction: Towards a New Philosophical-Psychoanalytic Materialism and Realism' in L. Chiesa (Ed.), *Lacan and Philosophy: The New Generation*, pp. 7–19. Melbourne: RePress.

Chiesa, L. (2016) *The Not-Two: Logic and God in Lacan*. Cambridge, MA: The MIT Press.

Lacan, J. (1998) *The Seminar of Jacques Lacan Book XX: Encore—On Feminine Sexuality, the Limits of Love and Knowledge 1972–1973*. London: W.W. Norton & Company.

Filmography

Spielberg, S. (Director) (2001) A.I. Artificial Intelligence [Film]. US: Warner Bros. Pictures.

Bibliography

Asimov, I. (2018) *The Complete Robot*. London: Harper Collins.

Auerbach, D. (2014) The Most Terrifying Thought Experiment of All Time. Available (01.03.2020) at: https://slate.com/technology/2014/07/rokos-basilisk-the-most-terrifying-thought-experimentof-all-time.html.

Ayerza, J. (2015) To Resume… *Lacanian Inc* 46: pp. 3–11.

Badiou, A. (2008) *Number and Numbers*. Cambridge: Polity Press.

Badiou, A. (2009) Antiphilosophy: Plato and Lacan. In A. Badiou (Ed.), *Conditions*. New York: Continuum.

Badiou, A. (2018) *Lacan: Anti-Philosophy 3*. New York: Columbia University Press.

Badiou, A., & Cassin, B. (2017) *There's No Such Thing as A Sexual Relationship: Two Lessons on Lacan*. New York: Columbia University Press.

Baudrillard, J. (1981) *Simulacra and Simulation*. Michigan: University of Michigan Press.

Baudrillard, J. (1988) *The Consumer Society: Myths and Structures*. Los Angeles: Sage Publications.

Baudrillard, J. (2000) *The Vital Illusion*. New York: Columbia University Press.

Baudrillard, J. (2005a) *The Intelligence of Evil or The Lucidity Pact*. London: Berg.

Baudrillard, J. (2005b) *The System of Objects*. London: Verso.

Baudrillard, J. (2008) *The Perfect Crime*. London: Verso.

Baudrillard, J. (2012) *The Ecstasy of Communication*. Los Angeles: Semiotext(e).

I. Millar, *The Psychoanalysis of Artificial Intelligence*, The Palgrave Lacan Series,
https://doi.org/10.1007/978-3-030-67981-1

Bostrom, N. (2014) *Superintelligence: Paths, Dangers, Strategies*. Oxford: Oxford University Press.

Bratton, B. (2015) 'Outing Artificial Intelligence: Reckoning with the Turing Test' in M. Pasquinelli (Ed.), *Augmented Intelligence and its Traumas*, pp. 69–80. Lüneburg: Meson Press.

Brenner, L. S. (2020) *The Autistic Subject: On the Threshold of Language*. London: Palgrave Macmillan.

Bristow, D. (2014) *2001: A Space Odyssey and Lacanian Psychoanalytic Theory*. London: Palgrave Macmillan.

Brooker, C., Jones, A., & Arnopp, J. (2018) *Inside Black Mirror*. London: Ebury Press.

Brousse, M-H. (2013) Ordinary Psychosis in the Light of Lacan's Theory of Discourse. *Psychoanalytic Notebooks* 26: pp. 23–31.

Butler, O.E. (2000) *Lilith's Brood*. New York: Warner Books.

Cahiers Kingston (2012) Synopsis of Jacques-Alain Miller, 'La Suture: Elements de la logique du signifiant'. Available (01.03.20) at: http://cahiers.kingston.ac.uk/synopses/syn1.3.html.

Canguilhem, G. (2016) What is Psychology? *Foucault Studies* 21: pp. 200–213.

Čapek, K. (2004) *R.U.R. (Rossum's Universal Robots)*. London: Penguin Books.

Chalmers, D. (2010) The Singularity: A Philosophical Analysis. *Journal of Consciousness Studies* 17(9–10): pp. 7–65.

Chiesa, L. (2007) *Subjectivity and Otherness: A Philosophical Reading of Lacan*. Cambridge, MA: The MIT Press.

Chiesa, L. (2014) 'Editorial Introduction: Towards a New Philosophical-Psychoanalytic Materialism and Realism' in L. Chiesa (Ed.), *Lacan and Philosophy: The New Generation*, pp. 7–19. Melbourne: RePress.

Chiesa, L. (2016) *The Not-Two: Logic and God in Lacan*. Cambridge, MA: The MIT Press.

Cixous, H. (1976) Fiction and its Phantoms: A Reading of Freud's Das Unheimliche (The "Uncanny"). *New Literary History* 7(3): pp. 525–548.

Copjec, J. (2015) *Read My Desire: Lacan Against the Historicists*. London: Verso.

De Halleux, B. (2016) Sexuality at the Time of the Speaking Body. *The Lacanian Review Hurly-Burly* 2: pp. 94–104.

Devlin, K. (2018) *Turned On: Science, Sex and Robots*. London: Bloomsbury.

De Sade, D.A.F. (2006) *Philosophy in the Boudoir*. London: Penguin.

Dolar, M. (1998) 'Cogito as the Subject of the Unconscious' in S. Žižek (Ed.), *Cogito and the Unconscious*, pp. 11–40. Durham: Duke University Press.

Dolar, M. (2006) *A Voice and Nothing More*. Cambridge, MA: The MIT Press.

Dreyfus, H. L. (1972) *What Computers Still Can't Do: A Critique of Artificial Reason*. Cambridge, MA: The MIT Press.

Dulsster, D. (2018) The Joke of Surplus Value and the Guffaw of the Saint. *Psychoanalytische Perspectieven* 36(2): pp. 205–224.

Edelman, L. (2004) *No Future: Queer Theory and the Death Drive*. Durham: Duke University Press.

EPFL (2020) Blue Brain Project. Available (01.03.20) at: https://www.epfl.ch/research/domains/bluebrain/.

Flisfeder, M. (2017) *Postmodern Theory and Blade Runner*. London: Verso.

Freud, S. (1898) 'Sexuality in the Aetiology of the Neurosis' in (2001) *The Standard Edition of the Complete Psychological Works of Sigmund Freud, Volume III (1893–1899): Early Psycho-Analytic Publications*, pp. 263–285. London: Vintage.

Freud, S. (1905) 'Fragment of an Analysis of a Case of Hysteria' in (2001) *The Standard Edition of the Complete Psychological Works of Sigmund Freud, Volume VII (1901–1905): A Case of Hysteria, Three Essays on Sexuality and Other Works*, pp. 7–123. London: Vintage.

Freud, S. (1911) 'Psycho-Analytic Notes on an Autobiographical Account of a Case of Paranoia (Dementia Paranoides)' in (2001) *The Standard Edition of the Complete Psychological Works of Sigmund Freud, Volume XII (1911–1913): Case History of Schreber, Papers on Technique and Other Works*, pp. 3–83. London: Vintage.

Freud, S. (1913) 'Totem and Taboo: Some Points of Agreement between the Mental Lives of Savages and Neurotics' in (2001) *The Standard Edition of the Complete Psychological Works of Sigmund Freud, Volume XIII*, pp. xiii–162. London: Vintage.

Freud, S. (1917) 'Introductory Lectures on Psycho-Analysis ' in (2001) *The Standard Edition of the Complete Psychological Works of Sigmund Freud, Volume XVI (1916–1917): Introductory Lectures on Psycho-Analysis (Part III)*, pp. 243–496. London: Vintage.

Freud, S. (1918) 'From the History of an Infantile Neurosis' in (2001) *The Standard Edition of the Complete Psychological Works of Sigmund Freud, Volume XVII (1917–1919): An Infantile Neurosis and Other Works*, pp. 1–124. London: Vintage.

Freud, S. (1919) 'The 'Uncanny" in (2001) *The Standard Edition of the Complete Psychological Works of Sigmund Freud, Volume XVII (1917–1919): An Infantile Neurosis and Other Works*, pp. 217–256. London: Vintage.

Freud, S. (2004) *Civilization and its Discontents*. London: Penguin.

Hammond, K. (2018) TedX: A New Philosophy on Artificial Intelligence [Video File]. Available (01.03.20) at: https://www.youtube.com/watch?v=tr9oe2TZiJw&t=179s.

Harari, Y.N. (2017) *Homo Deus: A Brief History of Tomorrow*. New York: HarperCollins Publishers.

Haraway, D. (2008) *When Species Meet*. Minneapolis: University of Minnesota Press.

Hauskeller, M. (2014) *Sex and The Posthuman Condition*. Basingstoke: Palgrave Macmillan.

Hester, H. (2018) *Xenofeminism*. Cambridge: Polity Press.

Johnson, D.K. (Ed.) (2020) *Black Mirror and Philosophy: Dark Reflections*. New Jersey: Wiley Blackwell.

Johnston, A. (2014) *Adventures in Transcendental Materialism: Dialogues with Contemporary Thinkers*. Edinburgh: Edinburgh University Press.

Johnston, A., & Malabou, C. (2013) *Self and Emotional Life: Philosophy, Psychoanalysis and Neuroscience*. New York: Columbia University Press.

Kang, M. (2017) The Mechanical Daughter of Rene Descartes: The Origin and History of an Intellectual Fable. *Modern Intellectual History* 14(3): pp. 633–660.

Kant, I. (1987) *Critique of Judgement*. Indianapolis: Hackett Publishing Company.

Kant, I. (1996) *Critique of Pure Reason*. Indianapolis: Hackett Publishing Company.

Kant, I. (2002) *Critique of Practical Reason*. Indianapolis: Hackett Publishing Company.

Kant, I. (2009) *Lectures on Logic*. Cambridge: Cambridge University Press.

Kant, I. (2007). *Anthropology, History, and Education*. Trans. M. Gregor, P. Guyer, R. Louden, H. Wilson, A. Wood, G. Zöller, and A. Zweig. New York: Cambridge University Press.

Kant, I. (2012) *Groundwork of the Metaphysics of Morals*. Cambridge: Cambridge University Press.

Kant, I. (2015) *Lectures on Anthropology*. Cambridge: Cambridge University Press.

Kittler, F. (1999) *Gramophone, Film, Typewriter*. Stanford: Stanford University Press.

Kittler, F. (2013) 'Flechsig, Schreber, Freud: An Information Network at the Turn of the Century' in F. Kittler (Ed.), *The Truth of the Technological World: Essays on the Genealogy of Presence*, pp. 57–68. California: Stanford University Press.

Kurzweil, R. (2005) *The Singularity is Near: When Humans Transcend Biology*. New York: Penguin.

Kurzweil, R. (2012) Foreword. In J. Neumann (Ed.), *The Computer and the Brain*. New Haven: Yale University Press.
Laboria Cuboniks (2008) *The Xenofeminist Manifesto: A Politics For Alienation*. London: Verso.
Lacan, J. (1977) *The Seminar of Jacques Lacan Book XI: The Four Fundamental Concepts of Psychoanalysis*. London: Karnac Books.
Lacan, J. (1988a) *The Seminar of Jacques Lacan Book II: The Ego in Freud's Theory and in the Technique of Psychoanalysis 1954–1955*. London: W.W. Norton & Company.
Lacan, J. (1988b) *The Seminar of Jacques Lacan Book I: Freud's Papers on Technique 1953–1954*. London: W.W. Norton & Company.
Lacan, J. (1990) *Television: A Challenge to the Psychoanalytic Establishment*: London: W.W. Norton & Company.
Lacan, J. (1992) *The Seminar of Jacques Lacan Book VII: The Ethics of Psychoanalysis*. London: W.W. Norton & Company.
Lacan, J. (1993) *The Seminar of Jacques Lacan Book III: The Psychosis 1955–1956*. London: Routledge.
Lacan, J. (1998) *The Seminar of Jacques Lacan Book XX: Encore—On Feminine Sexuality, the Limits of Love and Knowledge 1972–1973*. London: W.W. Norton & Company.
Lacan, J. (2001) 'Peut-être à Vincennes…' in J. Lacan (Ed.), *Autres écrits*, pp. 313–315. Paris: Éditions du Seuil.
Lacan, J. (2006a) 'Logical Time and the Assertion of Anticipated Certainty' in J. Lacan (Ed.), *Écrits*, pp. 161–175. London: W.W. Norton & Company.
Lacan, J. (2006b) 'The Function and Field of Speech and Language in Psychoanalysis' in *Écrits*, pp. 197–268. London: W.W. Norton & Company.
Lacan, J. (2006c) 'Science and Truth' in *Écrits*, pp. 726–745. London: W.W. Norton & Company.
Lacan, J. (2006d) 'Kant avec Sade' in *Écrits*, pp. 645–670. London: W.W. Norton & Company.
Lacan, J. (2006e) 'The Subversion of the Subject and the Dialectic of Desire in the Freudian Unconscious' in *Écrits*, pp. 671–702. London: W.W. Norton & Company.
Lacan, J. (2006f) 'On a Question Prior to Any Possible Treatment of Psychosis' in *Écrits*, pp. 445–488. London: W.W. Norton & Company.
Lacan, J. (2006g) 'The Mirror Stage as Formative of the *I* Function as Revealed in Psychoanalytic Experience' in *Écrits*, pp. 75–81. London: W.W. Norton & Company.

Lacan, J. (2006h) 'Position of the Unconscious' in *Écrits*, pp. 703–721. London: W.W. Norton & Company.

Lacan, J. (2007) *The Seminar of Jacques Lacan Book XVII: The Other Side of Psychoanalysis*. London: W.W. Norton & Company.

Lacan, J. (2009) L'Étourdit. *The Letter (Irish Journal for Lacanian Psychoanalysis)* 41: pp. 31–80.

Lacan, J. (2013) *On the Names-of-the-Father*. Cambridge: Polity Press.

Lacan, J. (2014) *The Seminar of Jacques Lacan Book X: Anxiety*. Cambridge: Polity Press.

Lacan, J. (2016) *The Seminar of Jacques Lacan Book XXIII: The sinthome*. Cambridge: Polity Press.

Lacan, J. (2018a) *The Seminar of Jacques Lacan Book XIX: …Or Worse*. Cambridge: Polity Press.

Lacan, J. (2018b) Note on the Child. *The Lacanian Review Hurly-Burly* 4: pp. 13–14.

Lacan, J. (2019) *The Seminar of Jacques Lacan Book VI: Desire and Its Interpretation*. Cambridge: Polity Press.

Land, N. (2011) *Fanged Noumena: Collected writings 1987–2007*. New York: Urbanomic.

Laurent, É. (2015) Gender and Jouissance. *Lacanian Ink* 46: pp. 66–89.

Laurent, É. (2016a) The Unconscious and the Body Event. *The Lacanian Review Hurly-Burly* 1: pp. 178–87.

Laurent, É. (2016b) *L'Envers de la biopolitique. Une ecriture pour la jouissance*. Paris: Navarin.

Lewis, S. (2019) *Full Surrogacy Now: Feminism Against Family*. London: Verso.

Liu, L.H. (2010) *The Freudian Robot: Digital Media and the Future of the Unconscious*. Chicago: The University of Chicago Press.

Lotringer, S. (1988) *Overexposed: Perverting Perversions*. Los Angeles: Semiotext(e).

Lovelock, J. (2019) *Novacene: The Coming Age of Hyperintelligene*. London: Penguin.

Malabou, C. (2008) *What Should We Do with Our Brain?* New York: Fordham University Press.

Malabou, C. (2012) *The New Wounded: From Neurosis to Brain Damage*. New York: Fordham University Press.

Malabou, C. (2019) *Morphing Intelligence: From IQ Measurement to Artificial Brains*. New York: Columbia University Press.

Maleval, J-C. (2012) Why the Hypothesis of an Autistic Structure? *Psychoanalytical Notebooks* 25.

Moravec, H. (1988) *Mind Children*. Cambridge, MA: Harvard University Press.

Mbembe, A. (2003) Necropolitics. *Public Culture* 15(1): pp. 11–40.

McGowan, T. (2007) *The Real Gaze: Film Theory After Lacan*. New York: SUNY Press.

McGowan, T. (2018) Like a Simile Instead of a Subject in Thakur, B. and Dickstein, J. (Eds.), *Lacan and the Nonhuman*. London: Palgrave Macmillan.

Merrin, W. (2005) *Baudrillard and the Media*. Cambridge: Polity Press.

Millar, I. (2018a) Black Mirror: From Lacan's Lathouse to Miller's Speaking Body. *Psychoanalytische Perspectieven* 36(2): pp. 187–204.

Millar, I. (2018b) Ex Machina: Sex, Knowledge and Artificial Intelligence. *Psychoanalytische Perspectieven* 36(4): pp. 447–467.

Millar, I. (2019) Kant avec Sade: A Ghost in the Shell? *Vestigia Journal* 2(1): pp. 154–172.

Millar, I. (2021) 'Before we even know what we are we fear to lose it: The Missing Object of the Primal Scene' in C. Neill (Ed.), *Blade Runner 2049: Some Lacanian Thoughts*. London: Palgrave Macmillan.

Miller, J-A. (1988) Extimité. *Prose Studies* 11(3): 121–31

Miller, J-A. (1990) 'Microscopia: An Introduction to the Reading of Television' in J. Lacan (Ed.), *Television: A Challenge to the Psychoanalytic Establishment*, pp. xi–xxxi. London: W.W Norton & Company.

Miller, J-A. (1998) Sobre "Kant con Sade". In *Elucidation de Lacan: Charles Brasileñas*. Buenos Aires: Paidos.

Miller, J-A. (2007a) 'Jacques Lacan and the Voice' in V. Voruz & B. Wolf (Eds.), *The later Lacan: An Introduction*, pp. 137–146. New York: SUNY.

Miller, J-A. (2007b) 'Interpretation in Reverse' in V. Voruz & B. Wolf (Eds.), *The Later Lacan: An Introduction*, pp. 3–9. New York: SUNY.

Miller, J-A. (2004) A Fantasy. Available (8th of April 2019) at: http://londonsociety-nls.org.uk/The-Laboratory-for-Lacanian-Politics/Some-Research-Resources/Miller_A-Fantasy.pdf.

Miller, J-A. (2012) 'Suture (Elements of the Logic of the Signifier)' in P. Hallward & K. Peden (Eds.), *Concept and Form Volume One: Key Texts from the Cahiers pour l'Analyse*, pp. 91–101. London: Verso.

Miller, J-A. (2013) 'You Are the Woman of the Other and I Desire You' Available (23.01.21) at https://www.lacan.com/essays/?page_id=331.

Miller, J-A. (2013a) The Real in the 21st century. *Hurly-Burly* 9: pp. 199–206.

Miller, J-A. (2013b) The Other Without Other. *Hurly-Burly* 10: pp. 15–29.

Miller, J-A. (2015a) The Unconscious and the Speaking Body. *Hurly-Burly* 12: pp. 119–132.

Miller, J-A. (2015b) Ordinary Psychosis Revisited. *Lacanian Ink* 46: pp. 90–115.

Miller, J-A. (2016) A New Alliance with Jouissance. *The Lacanian Review Hurly-Burly* 2: pp. 105–116.

Miller, J-A. (2019) Six Paradigms of Jouissance. *The Psychoanalytical Notebooks* 34: pp. 11–77.

Morel, G. (2006) The Sexual Sinthome. *Umbr(a)* 1: pp. 65–83.

Musk, E. (2018) Joe Rogan Podcast #1169 [Video File]. Available (01.03.20) at: https://www.youtube.com/watch?v=ycPr5-27vSI.

Neuralink (2019) Neuralink Launch Event [Video File]. Available (01.03.20) at: https://www.youtube.com/watch?v=r-vbh3t7WVI.

Nobus, D. (2019) 'Kant with Sade' in S. Vanheule, D. Hook & C. Neill (Eds.), *Reading Lacan's Écrits: From 'Signification of the Phallus' to 'Metaphor of the Subject'*, pp. 110–167. London: Routledge.

Parisi, L. (2015) 'Instrumental Reason, Algorithmic Capitalism and the Incomputable' in M. Pasquinelli (Ed.), *Alleys of Your Mind: Augmented Intelligence and its Traumas*, pp. 125–137. Lüneburg: Meson Press.

Pasquinelli, M. (2015) 'Introduction' in M. Pasquinelli (Ed.), *Alleys of Your Mind: Augmented Intelligence and its Traumas*, pp. 7–18. Lüneburg: Meson Press.

Pluth, E. (2019) 'Science and Truth' in S. Vanheule, D. Hook & C. Neill (Eds.), *Reading Lacan's Écrits: From 'Signification of the Phallus' to 'Metaphor of the Subject'*, pp. 268–307. London: Routledge.

Realbotix (2018) Projects. Available (01.03.20) on: https://realbotix.com/.

Richardson, K. (2018) Campaign Against Sex Robots. Available (01.03.20) on: https://campaignagainstsexrobots.org/.

Schreber, D.P. (2000) *Memoirs of My Nervous Illness*. Cambridge, MA: Harvard University Press.

Schwab, K. (2016) *The Fourth Industrial Revolution*. London: Penguin Random House.

Searle, J.R. (1980) Minds, Brains, and Programs. *Behavioral and Brain Sciences* 3(3): pp. 417–457.

Sharpe, M. (2015) Killing the Father, Parminedes: On Lacan's Antiphilosophy. *Continental Philosophy Review* 52: pp. 51–74.

Srnicek, N., & Williams, A. (2014) '#Accelerate Manifesto for an Accelerationist Politics' in N. Srnicek & A. Williams (Eds.) *#Accelerate: The Accelerationist Reader*, pp. 347–362. Falmouth: Urbanomic.

Stiegler, B. (1998) *Technics and Time 1: The Fault of Epimetheus*. Stanford: University Press.

Stiegler, B. (2013) *What Makes Life Worth Living?: On Pharmacology*. Cambridge: Polity Press.

Stiegler, B. (2014) *Symbolic Misery Volume 1: The Hyperindustrial Epoch*. Cambridge: Polity Press.

Tegmark, M. (2017) *Life 3.0: Being Human in the Age of Artificial Intelligence*. London: Penguin.

Teixera Pinto, A. (2018) The Psychology of Paranoid Irony. *Transmediale Journal* 1: pp. 18–22.

Tomšič, S. (2012) The Technology of Jouissance. *Umbr(a)* 17: pp. 143–158.

Tomsič, S. (2015) *The Capitalist Unconscious: Marx and Lacan*. London: Verso.

Tomšič, S. (2016) 'Psychoanalysis and Antiphilosophy: The Case of Jacques Lacan' in A. Cerda-Rueda (Ed.), *Sex and Nothing: Bridges from Psychoanalysis to Philosophy*, pp. 81–103. London: Karnac Books.

Turing, A. (1950) Computing Machinery and Intelligence. *Mind* 49: pp. 433–460.

Vanheule, S. (2014) *The Subject of Psychosis: A Lacanian Perspective*. Basingstoke: Palgrave Macmillan.

Vanheule, S. (2016) Capitalist Discourse, Subjectivity, and Lacanian Psychoanalysis. *Frontiers in Psychology* 7: pp. 1–14.

Vinge, V. (1993) The Coming Technological Singularity: How to Survive in the Post-Human Era. *Lewis Research Center, Vision 21: Interdisciplinary Science and Engineering in the Era of Cyberspace*: pp. 11–22.

Von Neumann, J. (2012) *The Computer and the Brain*. New Haven: Yale University Press.

Voruz, V. (2013) Disorder in the Real and Inexistence of the Other: What Subjective Effects? Available (01.03.20) at: www.iclo-nls.org/wp-content/uploads/Pdf/ICLO2013-VeroniqueVoruz.pdf.

Voruz, V. (2016) The Second Paternal Metaphor. Available (01.03.20) at: http://www.amp-nls.org/page/gb/49/nls-messager/0/2015-2016/2284.

Voruz, V. & Wolf, B. (2007) 'Preface' in V. Voruz & B. Wolf (Eds.), *The Later Lacan: An introduction*, pp. vii–xvii. New York: SUNY.

Wajcman, G. (2003) The Hysteric's Discourse. Available (01.03.20) at: https://www.lacan.com/hystericdiscf.htm.

Wolf, B. (2019) *Anxiety Between Desire and the Body: What Lacan Says in Seminar X*. London: Routledge.

Wright, C. (2018) 'Lacan's Cybernetic Theory of Causality: Repetition and the Unconscious in Duncan Jones' *Source Code*' in S. Matviyenko & J. Roof (Eds.), *Lacan and the Posthuman*, pp. 67–88. Basingstoke: Palgrave Macmillan.

Yudkowsky, E. (2010) *Timeless Decision Theory. The Machine Intelligence Research Institute*. Available (01.03.20) at: https://intelligence.org/files/TDT.pdf.

Žižek, S. (1993) *Tarrying with the Negative: Kant, Hegel and the Critique of Ideology*. Durham: Duke University Press.

Žižek, S. (1995) Woman is One of the Names of the Father, Or How Not to Misread Lacan's Formulas of Sexuation. Available (01.03.20) at: http://www.lacan.com/zizwoman.htm.

Žižek, S. (1997a) *The Plague of Fantasies*. London: Verso.

Žižek, S. (1997b) Desire: Drive = Truth: Knowledge. *Umbr(a)* 1: 147–151.

Žižek, S. (2005) *The Metastases of Enjoyment: On Women and Causality*. London: Verso.

Žižek, S. (2008) *For They Know Not What They Do: Enjoyment as a Political Factor*. London: Verso.

Žižek, S. (2016) *Disparities*. London: Bloomsbury.

Žižek, S. (2017a) *Incontinence of the Void: Economico-Philosophical Spandrels*. Massachusetts: MIT Press.

Žižek, S. (2017b) Blade Runner 2049: A View of Post-Human Capitalism. Available (01.03.20) at: https://thephilosophicalsalon.com/blade-runner-2049-view-of-post-human-capitalism/.

Žižek, S. (2020) *Sex and the Failed Absolute*: London: Bloomsbury.

Zupančič, A. (2000) *Ethics of the Real: Kant, Lacan*. London: Verso.

Zupančič, A. (2006) 'When Surplus Enjoyment Meets Surplus Value' in J. Clemens & R. Grigg (Eds.), *Reflections on Seminar XVII: Jacques Lacan and the Other Side of Psychoanalysis*, pp. 155–178. Durham: Duke University Press.

Zupančič, A. (2017) *What is Sex?* Cambridge, MA: MIT Press.

Filmography

Brooker, C. (Creator) (2011–) Black Mirror [TV Series]. UK: Channel 4.

Cameron, J. (Director) (1984) The Terminator [Film]. US: Orion Pictures.

Columbus, C. (Director) (1999) Bicentennial Man [Film]. US: Buena Vista Pictures.

Crichton, M. (Director) (1973) Westworld [Film]. US: Metro-Goldwyn-Mayer (MGM).

Foster, J. (Director) (2017) Black Mirror (S4E2): Arkangel [TV Series]. UK: Channel 4.

Garland, A. (Director) (2015) Ex Machina [Film]. US: Universal Pictures.

Jonze, S. (Director) (2013) Her [Film]. US: Warner Bros. Pictures.

Joy, L. & Nolan, J. (Creators) (2016–) Westworld [TV Series]. US: HBO.

Kleeman, J. (Creator) (2017) Rise of the Sex Robots [Documentary]. UK: The Guardian.

Kubrick, S. (Director) (1968) 2001: A Space Odyssey [Film]. US: Metro-Goldwyn-Mayer (MGM).

Pfister, W. (Director) (2014) Transcendence [Film]. US: Warner Bros Pictures.

Ridley, S. (Director) (1982) Blade Runner [Film]. US: Warner Bros. Pictures.

Sanders, R. (Director) (2017) Ghost in the Shell [Film]. US: Paramount Pictures.

Spielberg, S. (Director) (2001) A.I. Artificial Intelligence [Film]. US: Warner Bros. Pictures.

Villeneuve, D. (Director) (2017) Blade Runner 2049 [Film]. US: Warner Bros. Pictures.

Wachowski, L., & Wachowski, L. (Directors) (1999) The Matrix [Film]. US: Warner Bros. Pictures.

Index[1]

[1] Note: Page numbers followed by 'n' refer to notes.

I. Millar, *The Psychoanalysis of Artificial Intelligence*, The Palgrave Lacan Series,
https://doi.org/10.1007/978-3-030-67981-1

CPSIA information can be obtained
at www.ICGtesting.com
Printed in the USA
LVHW022229060423
743745LV00002B/127

9 783030 679804